Groundbreaking Scientific Experiments, Inventions and Discoveries of the 19th Century

GROUNDBREAKING SCIENTIFIC EXPERIMENTS, INVENTIONS AND DISCOVERIES OF THE 19TH CENTURY

MICHAEL WINDELSPECHT

ILLUSTRATED BY

SANDRA WINDELSPECHT

Groundbreaking Scientific Experiments, Inventions and
Discoveries through the Ages
ROBERT E. KREBS, SERIES ADVISER

GREENWOOD PRESS
Westport, Connecticut • London

Library of Congress Cataloging-in-Publication Data

Windelspecht, Michael, 1963–

    Groundbreaking scientific experiments, inventions, and discoveries of the 19th century / Michael Windelspecht.

       p. cm.—(Groundbreaking scientific experiments, inventions and discoveries through the ages)

    Includes bibliographical references and indexes.

    ISBN 0-313-31969-3 (alk. paper)

    1. Science—History—19th century.   2. Technology—History—19th century. I. Title. II. Series.

Q125.W79173   2003

509.034—dc21   2002075305

British Library Cataloguing in Publication Data is available.

Library of Congress Catalog Card Number: 2002075305

ISBN: 0-313-31969-3

First published in 2003

Greenwood Press, 88 Post Road West, Westport, CT 06881

An imprint of Greenwood Publishing Group, Inc.

www.greenwood.com

Printed in the United States of America

The paper used in this book complies with the Permanent Paper Standard issued by the National Information Standards Organization (Z39.48-1984).

10 9 8 7 6 5 4 3 2 1

For Dad
The apple never really does fall far from the tree

CONTENTS

LIST OF ENTRIES

SERIES FOREWORD

The material contained in five volumes in this series of historical ground-breaking experiments, inventions and discoveries encompasses many centuries from the prehistoric period up to the 20th century. Topics are explored from the time of prehistoric humans, the age of classical Greek and Roman science, the Christian era, the Middle Ages, the Renaissance period from the years 1350 to 1600, the beginnings of modern science of the 17th century, and great experiments, inventions, and discoveries of the 18th and 19th centuries. This historical approach to science by Greenwood Press is intended to provide students with the materials needed to examine science as a specialized discipline. The authors present the topics for each historical period alphabetically and include information about the women and men responsible for specific experiments, inventions, and discoveries.

All volumes concentrate on the physical and life sciences and follow the same historical format that describes the scientific developments of that period, In addition to the science of each historical period, the authors explore the implications of how historical groundbreaking experiments, inventions and discoveries influenced the thoughts and theories of future scientists and how these developments affected peoples' lives.

As readers progress through the volumes, it will become obvious that the nature of science is cumulative. In other words, scientists of one historical period draw upon and add to the ideas and theories of earlier periods. This is evident in contrast to the recent irrationalist philosophy of the history and philosophy of science that views science, not as a unique, self-correcting human empirical inductive activity, but as just another social or cultural activity where scientific knowledge is conjectural, scientific laws are contrived, scientific theories are all false, scientific facts are fickle, and scientific truths are relative. These volumes belie postmodern deconstructionist assertions that no scientific idea has greater validity than any other idea and that all "truths" are a matter of opinion.

For example, in 1992 the plurality opinion by three jurists of the U.S. Supreme Court in *Planned Parenthood v. Case* restated the "right" to abortion by stating: "*at the heart of liberty is the right to define one's own concept of existence, of meaning of the universe, and of the mystery of human life.*" This is a remarkable deconstructionist statement, not because it supports the right to abortion, but because the Court supports the relativistic premise that anyone's concept of the universe is whatever that person wants it to be, and not what the universe actually is based on: what science has determined by experimentation, the use of statistical probabilities, and empirical inductive logic.

When scientists develop factual knowledge as to the nature of nature they understand that "rational assurance is not the same thing as perfect certainty." By applying statistical probability to new factual data this knowledge provides the basis for building scientific hypotheses, theories and laws over time. Thus, scientific knowledge becomes self-correcting as well as cumulative.

In addition, this series refutes the claim that each historical theory is based on a false paradigm (a methodological framework) that is discarded and later is just superseded by a new, more recent theory also based on a false paradigm. Scientific knowledge is the sequential nature that revises, adds to and builds on old ideas and theories as new theories are developed based on new knowledge.

Astronomy is a prime example of how science progressed over the centuries. Lives of people who lived in the prehistorical period were geared to the movement of the sun, moon and stars. Cultures in all countries developed many rituals based on observations of how nature affected the flow of life, including the female menstrual cycle and people's migrations to follow food supplies or adaptations to survive harsh winters. Later, after the discovery of agriculture around 8000 to 9000 B.C.E., people learned to relate climate and weather, the phases of the moon and the periodicity of the sun's apparent motion in relation to the Earth, because these astronomical phenomena seemed to determine the fate of their crops.

The invention of bronze by alloying first arsenic and later tin with copper occurred about 3000 B.C.E. Much later, after discovering how to use the iron found in celestial meteorites, and still later, in 1000 B.C.E. when people learned how to smelt iron from ore, civilization entered the Iron Age. The people of the Tigris-Euphrates region invented the first calendar based on the phases of the moon and seasons in about 2800 B.C.E. During the ancient and classical Greek and Roman periods (about 700 B.C.E. to A.D. 100) mythical gods were devised to explain what was viewed in the heavens or to justify their behavior. Myths based on astronomy, such as the sun and planet gods as well as Gaia the Earth mother, were part of their religions and affected their way of life. This period was the beginning of the philosophical thoughts of Aristotle and others concerning astronomy and nature in general that pre-dated modern science. In about 235 B.C.E. the Greeks first proposed a helio-centric relationship of the sun and planets. Ancient people in Asia, Egypt and India invented fantastic structures to assist the unaided eye in viewing the posi-tions and motions of the moon, stars and sun. These instruments were the forerunners of the modern telescopes and other devices that make modern astronomical discoveries possible. Ancient astrology was based on the belief that the positions of bodies in the heavens controlled one's life. Astrology is still confused with the science of astronomy and is still not based on any reli-able astronomical data.

The ancients knew that a dewdrop on a leaf seemed to magnify the leaf's surface. This led to the invention of a glass bead that could be used as a magnifying glass. In 1590 Zacharias Janssen, an eyeglass maker, dis-covered that two convex lenses, one at each end of a tube, increased the magnification. In 1608 Hans Lippershey's assistant turned the instrument end-to-end and discovered that distant objects appeared closer; thus the telescope was discovered. The telescope has been used for both navigation and astronomical observations since the 17th century. The invention of new instruments, such as the microscope and the telescope, led to further discov-eries such as of the cell by Robert Hooke and the four moons of Jupiter by Galileo, who made this important astronomical discovery that revolutionized astronomy with a telescope of his own design and construction. These inven-tions and discoveries enabled the expansion of astronomy from an ancient "eyeball" science to an ever-expanding series of experiments and discoveries leading to many new theories about the universe. Others invented and improved astronomical instruments, such as the reflecting telescope combined with photography, the spectroscope, and Earth-orbiting astronomical instru-ments that resulted in the discovery of new planets and galaxies as well as new theories related to astronomy and the universe in the 20th century. The age of "enlightenment" through the 18th and 19th centuries culminated in an explosion of new knowledge of the universe that continued through the 20th and into the 21st centuries. Scientific laws, theories, and facts we now know about astronomy and the universe are grounded in the experiments, discov-eries and inventions of the past centuries, just as they are in all areas of science.

The books in the series Groundbreaking Experiments, Inventions and Dis-coveries, are written in easy-to-understand language with a minimum of scientific jargon. They are appropriate references for middle and senior high school audiences as well as for the college-level nonscience major and for the general public interested in the development and progression of science over the ages.

Robert E. Krebs
Series Adviser

ACKNOWLEDGMENTS

I would like to recognize the assistance of several individuals who made substantial contributions toward this volume of the series. First I would like to thank Dr. Robert Krebs, the series editor, for providing me with the opportunity to contribute a second volume in the series. I have held a lifelong interest in the history of science, and Dr. Krebs has given me the opportunity to pass that interest on to the next generation of readers. Dr. Krebs has also provided insight and invaluable comments for many of the entries that have greatly enriched the content of the work. Next I would like to thank my editors, Emily Birch and Debby Adams, and the staff of Greenwood Publishing, for their assistance through the entire process.

A significant amount of the artwork in this volume was produced by Sandra Windelspecht, who as my wife held the double duty of being both an illustrator and a support system for this volume. Her design of the artwork in many cases was vastly superior to anything that I had originally envisioned, and I believe that this greatly enhances the presentation of the material. Additional artwork was provided by the National Library of Medicine, the Library of Congress, and Hulton Archive Images. The staff at these organizations were, as always, exceptionally helpful in providing assistance with my requests.

As with the previous volume that I contributed, I would like to thank Greenwood Publishing for recognizing the importance of producing this series. I frequently find as an educator that students are fascinated by the history of science and the lives of the people who are responsible for our modern world. In many cases severe misconceptions about the history of science exist, and I hope that works such as these assist people in understanding the relationship between science and society. Science is not a tool for scientists but a gift for understanding the natural world in which we all live.

INTRODUCTION

The importance of the 19th century to the history of science and technology is best represented by the names that are frequently assigned to this period. In mathematics this is commonly called the "Golden Age of Mathematics." Many historians consider the 19th century, especially the latter half, to be the start of the modern era of science, because many of our current ideas and theories of the natural world were initiated during this time. Just as the scientific community experienced a Scientific Revolution in the 17th century, the 19th century is frequently referred to as the center of the Industrial Revolution, when technology and inventions forever changed society. Each of these instances signify that important events occurred during the period, but in reality the 19th century is actually a culmination of almost two centuries of work by scientists and mathematicians.

The Scientific Revolution of the 17th century forever changed the way in which scientists view the natural and physical world. In the place of the method of the ancient Greeks, which involved observation followed primarily by philosophical discussion, arose a method of experimentation and validation that persists to this day. The science of the 17th century also introduced the formation of scientific societies and journals in which scientists presented their ideas and results to be critiqued by their peers. During the 17th century mathematics, astronomy and the physical sciences flourished, and names like Newton, Kepler and Galileo dominated the century. This is not to say that important advances were not made in the biological and chemical sciences, but in many cases these were still descriptive. During the 18th century chemistry as a science was finally separated from the study of alchemy by the action of scientists such as Lavoisier. Also during this time the naturalists, the early name given to biologists, began to study the structure of the natural world and to question the processes by which it functioned and changed over time.

In the 19th century these ideas and practices culminated in some of the most important theories in the history of science.

The biological sciences became an important force in the study of the natural world during the 19th century. Previously the study of living creatures had been primarily descriptive as naturalists collected samples and attempted to establish order to the natural world in the same manner as the chemists and physicists did in their fields. Early naturalists recognized that organisms changed over time but lacked an explanation for what was causing the process. Mechanisms of change had to be developed by early evolutionary theorists to account for the relatively rapid changes in organisms on an Earth that was very young according to biblical scholars. However, advances in the study of geology, most notably by Charles Lyell, forever changed the concept of geological time and established the framework for one of the more important theories of the 19th century. In 1859 Charles Darwin published his ideas on how organisms change over time in *The Origin of Species*. This landmark publication revolutionized the study of the biological sciences and established one of the greatest controversies of the modern age, the question over divine creation or common descent as the process of human evolution.

The 19th century was also a time for new ideas in the study of medicine. Preventive medicine, a major component of modern healthcare systems, originated in the 1800s as a result of the development of vaccines and the realization that diseases may be caused by microscopic organisms. The work of Louis Pasteur ensured that generations of humans would be protected from the natural microbe content of food supplies from his simple experiments with a process that is now called pasteurization. Other advances, such as the discovery of synthetic dyes, were the beginning of the modern practice of studying diseases using technology. Another development in the study of medicine was the invention of specialized instruments, which advanced the specialization of physicians into diverse areas of medical practice in the 20th century.

The influence of the scientific advances of the 19th century on modern science and society cannot be underestimated. The term *science* itself is a 19th century creation, first used in 1851 by the English philosopher William Whewell. The word is derived from the Latin word *scientia* ("knowing"), and it accurately reflects what the goal of 19th century scientific studies was. The scientists of the 19th century were dedicated not only to describing the natural world but also to knowing how it functioned and was structured. To do this scientists and inventors participated in more extensive research and professional organizations, and also developed research laboratories dedicated to the solving of specific problems. The most famous of these is probably the one Thomas Edison founded at Menlo Park, New Jersey. At this site Edison applied scientific principles to the development of many inventions, including the incandescent light bulb, and established some of the principles of modern electric power, which in turn shaped the structure of 20th century society.

The relationship between scientific studies and technological advances was recognized and firmly established in the 19th century. In many cases this was a reciprocal arrangement: inventors sometimes provided the instruments for advanced scientific studies, and in other cases the scientists providing the basic science that fostered the ideas of the inventors. An invention or discovery frequently resulted in a cascade of events. For example, the study of steam engines led to investigations into the study of thermodynamics, which in turn disproved many of the existing concepts regarding heat. The design of improved steam engines not only powered the Industrial Revolution, but also enabled the western expansion of the United States, which interested inventors in designing a means of communicating across greater distances. The invention of the telegraph and telephone soon followed.

Another important aspect of the 19th century was the decentralization of the scientific community. For most of the history of science, isolated hotspots of scientific inquiry have existed. Early in the history of science scientists congregated for study and support in small areas of ancient Greece and Alexandria, Egypt, for example. Areas of study existed in China, the Arab world, and India even before the fall of the Roman Empire. During the Scientific Revolution the majority of the scientific advances occurred in Europe. By the 19th century, though, the study of science became more global. Europe continued to play a major role, as it had for several centuries, but scientific communities were developing in the United States and Russia. The accomplishments of Russian and American scientists would dominate much of the science of the early 20th century. This was the beginning of a trend that continues to this day, and scientific and technological advances occur on every continent of the globe. Of course decentralization could also inhibit the exchange of information, but by the late 19th century inventions such as the telegraph and telephone, as well as improvements in means of transportation such as the steam engine and internal combustion engine, would ensure that the scientists of the 20th century could easily be in contact with colleagues across the world. The modern Internet, originally constructed for military use, is an extension of these technologies and has become an importance avenue by which the scientific community communicates its findings in the electronic world.

This book is designed as a reference volume for anyone interested in obtaining an overview of the advances in mathematics, science and technology during the 19th century. The topics in this volume were chosen from a historical perspective because of their influence on the development of science, mathematics and technology. By no means is this a comprehensive compilation of all of the accomplishments during this time frame. Entries were chosen not only because of their influence, but also to indicate the wide array of accomplishments during this century. In each case, though, the work of these investigators eventually led to scientific, mathematical or technological breakthroughs that shaped the course of science, and often society, in the following centuries. Each entry provides a brief history of the topic, which in some cases dates from the time of the ancient Greeks but in others goes back only a few decades. This allows the reader to more fully understand the climate of the times and thus appreciate the methodology by which the scientists and inventors approached their discoveries. Because many advances of the 19th century were built on the achievements of many disciplines, each entry is cross-referenced to other topics in the book that may provide additional insight on related events that occurred during this time. Each entry contains descriptions of experiments, discoveries, and inventions and how the specific achievement influenced the science and culture of the times. Although, described from the perspective of 19th century science and culture, some of these creations were so revolutionary that their potential influence on science and society were not completely recognized during the time frame of the 19th century. However, all of the chosen items made important contributions to the development of modern science and society in the 20th and 21st centuries.

This book is targeted for the general science audience and I have tried to use common language to describe many of the scientific and mathematical discoveries. For areas that require a deeper understanding of technical and scientific terminology, a Glossary appears at the end of the book. Each entry contained in the Glossary is highlighted in boldface type when it is first used in the text. Also included for each entry is a bibliography of reference materials that direct the reader to sources of additional information, and a complete Bibliography is included. An index of subjects and names used within the work allow readers to quickly access desired information from the century covered by this book. The timeline at the front of the book helps to illustrate other advances that were occurring during the century. These reference materials make this volume attractive for secondary school libraries as well as for undergraduate students in colleges and universities, where students may be seeking general information about a specific experiment or discovery. In addition, community libraries that wish to possess a general reference volume on the 19th century, as well as anyone with an interest in science history, will find this work a useful addition to their collection.

TIMELINE OF
IMPORTANT EVENTS

1796   Edward Jenner develops a vaccine against smallpox.

1799   Friedrich Gauss proposes the fundamental theory of algebra.

1800   William Herschel discovers the infrared portion of the electromagnetic spectrum.

1800   Alessandro Volta invents the first electric cell, the precursor to the modern electric battery.

1801   Thomas Young begins to challenge the existing philosophy that light exists as a particle.

1801   Carl Gauss publishes his ideas on number theory, which influences much of the remaining 19th century studies of mathematics.

1801   William Ritter discovers ultraviolet radiation.

1801   Giuseppe Piazzi discovers the first asteroid and names it Ceres.

1803   John Dalton develops the theory of definite proportions to explain the atomic basis of compounds.

1803   Richard Trevithick invents the first steam locomotive.

1807   Using electrochemistry Humphry Davy discovers the elements potassium and sodium.

1808   John Dalton publishes *New System of Chemical Philosophy*, which introduces the modern atomic theory to science.

1808   Gay-Lussac develops the law of combining volumes for gases.

1809   William Wollaston invents the reflecting goniometer, which is used extensively in the study of mineralogy.

1809   Jean Baptiste Lamarck publishes his theory of acquired characteristics.

1810   John Dalton presents one of the first tables of atomic weights.

1811   Avogadro begins development of several important gas laws, including Avogadro's hypothesis and Avogadro's law.

1811   Friedrich Koenig introduces the first successful version of the cylinder printing press.

1812   Georges Cuvier publishes his theory of catastrophism as a mechanism for explaining the fossil record.

1813   Jöns Berzelius develops a system of chemical symbols that denotes individual elements in a compound using the first letter of the element's name.

1814   Joseph von Fraunhofer performs the first detailed spectral analysis.

1816   William Smith publishes *Strata Identified by Organized Fossils*, which establishes the basis for using fossils as a chronological reference.

1816   René Laennec invents the first stethoscope.

1818   The work of Thomas Young, Augustin Fresnel and Francois Arago proves that light is a wave.

1819   The Dulong-Petit Law is presented and shows the relationship between specific heat and atomic weight.

1819   Hans Oersted provides the first experimental evidence of the link between electricity and magnetism.

1821   Michael Faraday discovers the process of electromagnetic induction.

1822   Joseph Nicéphore Niépce produces some of the first photographic images, which are called heliographs.

1823   Justus von Liebig and Friedrich Wohler independently discover isomers in chemistry.

1823   William Sturgeon invents the first electromagnet.

1824   Nicolas Carnot publishes *On the Motive Power of Fire*, which influences the development of the laws of thermodynamics and improvements to the steam engine.

1827   Georg Ohm describes the mathematical relationship among current, resistance and voltage in Ohm's law.

1829   Nikolay Lobachevsky proposes the first non-Euclidian geometry.

1830   Charles Lyell publishes *The Principles of Geology*, which has a strong influence on the development of both geological and evolutionary theories over the next few decades.

1831   Michael Faraday invents the first electrical generator, and shortly afterward Joseph Henry designs the first electric motor.

1831   Robert Brown writes detailed descriptions of the cell nucleus in plants.

1832   Charles Babbage invents the difference engine, the precursor of the modern computer.

1833   Anselme Payen and Jean Persoz isolate the first enzyme and name it diastase.

1834   George Boole establishes the principles of what will eventually be called Boolean algebra.

1837   Edward Davy invents a method of using a small battery to enhance an electrical signal.

1837   Pierre Wantzel provides the proof for the Euclidian geometry idea of trisecting an angle.

1837   Joseph Henry invents the electric relay.

1837   Louis Jacques Daguerre produces the daguerreotype, an improved method of producing photographic images.

1837   Henri Joachim Dutrochet demonstrates that chlorophyll is necessary for photosynthesis.

1838   Mathias Schleiden publishes his views on the cellular basis of life in plants.

1838   Samuel Morse invents the Morse code for his version of the telegraph.

1838   Friedrich Bessel makes the first accurate estimate of stellar distances when he calculates the distance to 61 Cygni to be approximately 11 light-years.

1839   Johannes Mulder develops the term *protein* to describe biological molecules containing nitrogen.

1839   Charles Goodyear develops vulcanized rubber.

1839   Theodor Schwann presents his ideas on the cell theory, thus uniting all life at the cellular level.

1839   John Draper makes the first use of photography in astronomy when he photographs Earth's moon.

1842   Crawford Long documents the first use of ether as an anesthetic.

1842   Christian Doppler discovers the Doppler effect, which plays an important role in astronomical studies.

1843   William Rankine presents his ideas on the conservation of mechanical energy.

1843   James Prescott Joule begins his studies of heat, which lead to the development of the mechanical equivalent of heat in 1847.

1843   William Hamilton develops quaternions to explain the algebraic treatment of complex numbers in four dimensions.

1845   Michael Faraday demonstrates that an electromagnetic wave can influence the polarization of light. This is called the Faraday effect.

1845	Robert Remak defines the role of each of the cell layers of embryonic development.
1845	Christian Shonbein discovers the explosive chemical nitrocellulose, also called guncotton.
1846	Based on the calculations of Joseph Le Verrier and John Couch Adams, Gottfried Galle discovers the planet Neptune.
1847	James Simpson uses chloroform to reduce the pain of childbirth.
1847	Hermann von Helmholtz makes the first formal presentation of the first law of thermodynamics.
1847	Ascanio Sobrero invents nitroglycerine.
1848	William Thomson calculates the value for absolute zero.
1849	The first underwater telegraph cable is installed beneath the Connecticut River.
1850	Rudolf Clausius presents the second law of thermodynamics.
1850	Hermann von Helmholtz invents the first ophthalmoscope.
1851	Jean Foucault experimentally demonstrates that the Earth rotates on its axis.
1852	Edward Frankland identifies the combining power of organic compounds, establishing the foundation for the concept of valency.
1852	John Lamont and Edward Sabine discover that sunspot activity varies the intensity of the Earth's magnetic field.
1852	Elisha Graves Otis invents a safety device for elevators that greatly increases their popularity by the end of the century.
1853	Abraham Gesner develops a way to refine kerosene.
1854	Hermann von Helmholtz calculates that the Sun is 25 million years old.
1854	Johann Geissler constructs the earliest form of a cathode-ray tube.
1856	Norman Pogson develops the mathematical basis for determining the magnitude (brightness) of a star.
1856	The first fossil skull of a Neanderthal is discovered.
1856	William Perkin manufactures the first synthetic dye.
1856	Henry Bessemer invents the blast furnace for steel production.
1858	Archibald Couper determines that carbon is a tetravalent element.
1859	James Clerk Maxwell and Ludwig Boltzmann independently develop the kinetic theory of gases.
1859	Edwin Drake drills the first oil well in Titusville, Pennsylvania.
1859	Charles Darwin publishes *On the Origin of Species*.
1860	Hermann Kolbe determines the chemical structure of salicylic acid, the active component of aspirin.

1877	Thomas Edison patents the first phonograph that records and plays back sound.
1880	John Milne invents a form of seismometer that is still frequently used today.
1880	Werner von Siemens invents the first electric elevator.
1881	Thomas Edison, Joseph Swan and others invent the first incandescent lamp.
1881	Albert Michelson invents the interferometer, which is used to disprove the existence of ether.
1882	Pi is shown to be transcendental by Ferdinand Lindemann.
1882	Thomas Edison begins electricity service to areas of New York City.
1883	James Hall develops the idea of geosynclines to explain the process of mountain formation.
1884	Charles Parsons invents the modern steam turbine.
1884	Carl Koller uses cocaine as a local anesthetic.
1884	Ottmar Merganthaler invents the linotype, a device that increases the efficiency of the printing industry.
1885	William Stanley invents a transformer that steps voltage down or up.
1885	Loius Pasteur develops a vaccine against rabies.
1885	Karl Benz invents the first commercially available automobiles.
1887	Rudolf Hertz makes the first calculation of a wavelength.
1888	Nikola Tesla invents an electric motor that efficiently uses alternating current.
1888	George Eastman invents the Kodak camera, which makes use of the newly invented celluloid film.
1890	Emil von Behring invents a procedure to use antitoxins for vaccinations.
1891	Max Wolf uses photography to discover the asteroid Brucia. Over the next several decades a large number of asteroids are discovered using this procedure.
1892	William Seward Burroughs invents a calculating machine that uses a keypad to input data.
1893	Rudolf Hertz's book *Electric Waves* establishes the principles of radio communication.
1895	William Röntgen discovers x-rays.
1896	Antoine-Henri Becquerel discovers alpha and beta radiation, which are further classified later the same year by Ernest Rutherford.
1897	Guglielmo Marconi invents the first radio capable of sending wireless communications over long distances.

1897   Joseph John Thomson discovers the electron.

1897   Hans and Eduard Buchner demonstrate that enzymatic reactions can be conducted in vitro.

1897   Rudolf Karl Diesel designs a successful internal combustion engine that uses compression to ignite the fuel.

1898   Marie Curie discovers polonium while investigating the radioactive nature of uranium. This is the beginning of the study of radiochemistry.

1898   The first near-Earth asteroid, named Eros, is discovered by Gustav Witt.

1898   James Dewar succeeds in liquefying hydrogen.

1899   David Hilbert establishes one of the standard of axioms for geometric studies.

1900   Paul Villard discovers gamma rays.

1900   Hugo de Vries, Carl Correns and Erich von Tschermak independently rediscover Mendel's principles of inheritance.

1900   Max Planck marks the beginning of modern physics with his explanation of quantum physics.

1901   Emil Fischer demonstrates that proteins are formed from amino acids, and later artificially constructs a protein in the lab.

1901   Guglielmo Marconi successfully transmits a radio signal across the Atlantic Ocean.

1906   William Nernst develops the third law of thermodynamics.

EXPERIMENTS, INVENTIONS AND DISCOVERIES

A

Absolute Zero (1848–1899): The 19th century was a time of many important advances in the study of the physical sciences. One of the leading areas of investigation involved understanding the physical properties of heat. At the start of the century scientists had incorrectly identified heat as a physical substance called caloric that was believed to be transferred between compounds during a chemical reaction. However, by the mid-19th century scientists recognized that heat was actually the result of the motion of atoms at the atomic level (see HEAT). During the same period physicists were beginning to understand that the temperature of an object corresponded to the average kinetic motion of its atoms. The majority of this early work on heat was performed with gases, and much of it was summarized in 1860 with the development of the kinetic theory of gases (see KINETIC THEORY OF GASES).

If temperature is a measure of the motion of atoms, then theoretically as temperature decreases at some point all **molecular** motion should cease. Because this point represents the lowest possible temperature, it is called absolute zero. The study of absolute zero in the 19th century was the culmination of a number of important studies. Since the 17th century scientists had recognized that a relationship existed between the temperature and volume of a gas, and during the 19th century a number of studies were done on the relationship between the temperature of a gas and the pressure in a container (see GAS LAWS). In addition, by the mid-19th century physicists had been working to define the laws of thermodynamics (see LAWS OF THERMODYNAMICS). In many cases these scientists wished to study the properties of heat independent of the state of **matter**. The existing temperature scales of the time, Celsius and Fahrenheit, were convenient for the measuring of heat as it related to the physical states of water. What was needed was a temperature scale dedicated to the measurement of heat.

Scottish scientist William Thomson (1824–1907) is frequently given credit for the development of this temperature scale, although there is evidence that a number of scientists were working on similar ideas around the same time. Thomson, also known as Lord Kelvin, was interested in developing a temperature scale for measuring the **absolute temperature** of matter as part of his work on **entropy**, or what was later called the second law of thermodynamics (see LAWS OF THERMODYNAMICS). As with all temperature scales, a reference point was needed from which the remainder of scale could be constructed. As a reference point, Thomson chose the triple point of water. The triple point of water, equal to 0.01°C, is the temperature at which water simultaneously exists as a gas, a liquid and a solid. From the study of gas laws (see GAS LAWS), Thomson recognized that the pressure of a gas decreases as the temperature is lowered. This is caused by a lessening of the motion of the gas molecules as defined by the kinetic theory of gases. By extrapolating backwards mathematically, Thomson was able to calculate the temperature at which all molecular motion ceased, or absolute zero, at −273°C. This temperature scale, based on the value of absolute zero, is called the Kelvin scale in honor of Thomson's work. Temperatures on the Kelvin scale are indicated by the letter "K" (without a degree sign) to distinguish them from the other scales. Figure 1 gives important reference points of the three common temperature scales in use today.

As the 19th century drew to a close, physicists had become increasingly interested in the liquefaction of gases using high pressures and low temperatures. The studies using pressure were basically an extension of Boyle's law, which states that a direct relationship exists between pressure and volume. However, this process alone did not work for all elements. In 1895 the German chemist Carl Gottfried von Linde (1842–1934) invented a method of cooling gases. To achieve the liquefaction of some elements, specifically hydrogen, it became necessary to approach temperatures near absolute zero. In 1898 the English chemist James Dewar (1842–1923) liquefied hydrogen at temperatures around 20°K. The following year Dewar succeeded in solidifying hydrogen at a temperature of 14°K. These represent the closest approaches to absolute zero in the 19th century.

The study of absolute zero in the 19th century represents an important achievement for the science of that time, but it had a far greater impact on the study of the physical sciences in the 20th century. In 1906, the German scientist Walther Nernst (1864–1941) developed the third law of thermodynamics, which defines entropy at temperatures close to absolute zero. He was awarded the 1920 Nobel Prize in chemistry for this work. The theoretical value of absolute zero, recognized to be −273.16°C, represented a challenge to physicists, and many attempted to see how close they could get to this value.

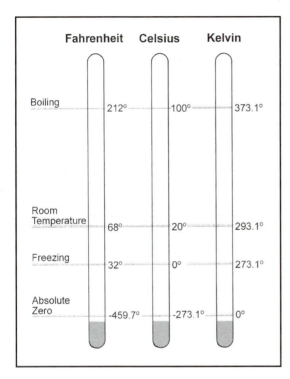

Figure 1. A comparison of the Fahrenheit, Celsius and Kelvin temperature scales indicating the principal reference points of 19th century studies.

During the 20th century physicists used a variety of new techniques in an attempt to approach absolute zero. For example, in 1933 the American chemist William Giauque (1895–1982) used magnetism to supercool a compound to 0.25 K. This type of technique became a popular approach for the study of absolute zero, and for his work Giauque was awarded the 1949 Nobel Prize in chemistry. In 1962, using a new technique involving the mixing of **isotopes**, the English physicist Heinz London (1907–1970) achieved a temperature 1 millionth of a degree above absolute zero.

The study of absolute zero continues into modern science today. However, most of the research in this area is now directed at achieving ultralow temperatures for the study of superconductivity, specifically as it relates to computer and information system technology, since supercool conductors have a greater capacity for transferring data. In 1972, the American scientists John Bardeen (1908–1991), Leon Cooper (1930–) and John Schrieffer (1931–) received the Noble Prize in physics, in part for their work on superconductivity at temperatures approaching absolute zero. In 1997, Steven Chu (1948–), Claude Cohen-Tannoudji (1933–) and William Phillips (1948–) received the Nobel Prize in physics for the manipulation of atoms at temperatures 240 millionths of a degree above absolute zero. Although theoretical attempts have been made to achieve temperatures below absolute zero, the boundaries established by William Thomson in the 19th century obviously continue to challenge physicists and find application in modern science.

Selected Bibliography

Asimov, Isaac. *Asimov's New Guide to Science*. New York: Basic Books, 1984.

Krebs, Robert E. *Scientific Laws, Principles and Theories: A Reference Guide*. Westport, CT: Greenwood Press, 2001.

Purrington, Robert D. *Physics in the Nineteenth Century*. New Brunswick, NJ: Rutgers University Press, 1997.

Algebra (1799–1854): In its simplest form algebra is the branch of mathematics that involves the solving of equations. For much of the history of science, mathematicians have studied the properties of equations of a degree higher than one, meaning that one of the variables has associated with it exponential notation, such as $x^2 + 2y = 7$. This equation would be considered a second-degree equation based on the exponent assigned to the variable x. The Chinese and Arabic cultures were well versed in the study of algebra, and until the Scientific Revolution in Europe in many cases they were ahead of their Western counterparts in most aspects of algebraic studies. In Europe during the 17th century the application of algebraic principles to geometry began with the work of Pierre de Fermat (1601–1665) and René Descartes (1596–1650), two French mathematicians. This form of mathematics, also called **analytical geometry**, established the foundations not only for the development of **calculus**, but also for a series of advances in the study of geometry in the 19th century (see GEOMETRY). However, the study of algebra was not limited to geometric shapes, and several important advances were made in the study of equations during the 19th century that played an influential role in the 20th century.

One of the leading mathematicians of the early 19th century was the German mathematician Carl Friedrich Gauss (1777–1855). Gauss played an important role in defining the properties of complex numbers, which are frequently used to study the square roots of negative numbers (see NUMBERS AND NUMBER THEORY). One of Gauss's greatest contributions to the study of algebra was the introduction of the fundamental theory of algebra (1799). For several centuries mathematicians had been attempting to devise mechanisms to determine the number of roots (or solutions) of a polynomial of a degree higher than three. A polynomial is a mathematical equation with multiple terms, for example $x^3 + 2x^2 - 13 = 4$ (a third-degree equation). In the

16th century the Italian mathematicians Ludovico Ferrari (1522–1565) and Niccolo Tartaglia (ca. 1500–1557) had developed the properties for solving cubic equations, those with a degree of three, and quartic equations, those with a degree of four. However, as the degree of the equation increases, determining all of the roots of the equation becomes significantly more difficult. Gauss's theory established that for a single variable equation of degree n, there are n roots. Gauss also developed several important principles for dealing with equations in two variables. His work was not limited to the study of algebra, as he also played an important role in applying mathematics to astronomy, where he assisted 19th century astronomers in the identification of asteroids (see ASTEROIDS).

Another important mathematician of the early 19th century was Evariste Galois (1811–1832). During his impressive, but short career Galois developed what is commonly called the group theory. The group theory is an expansion on the work of several earlier mathematicians. In the 18th century French mathematician Joseph Louis Lagrange (1736–1813) investigated solving polynomial equations using a process called factorization. Factorization involves finding a common multiple of the equation that can be removed from each term of the polynomial to reduce the overall degree of the polynomial. The principles for this were first established in the 17th century by Descartes and greatly expanded by Lagrange. Lagrange focused on using permutations, or variations of the order of the terms, as a process for solving the roots of a polynomial. This work was further explored by the French mathematician Augustin-Louis Cauchy (1789–1857) in the early 19th century (ca. 1815). Galois suggested that sets of permutations could be grouped together and treated as a single part of solving an equation. Galois called this the group of the equation, and it became an important contribution in developing an abstract method to solve high-order algebraic equations. Several years later Gauss used the group theory to explain quadratic equations, or those of a degree no higher than two.

Equations of the fifth degree (x^5), also called quintic equations, had presented a problem for mathematicians for some time. At the very end of the 18th century the Italian mathematician Paolo Ruffini (1765–1822) presented a proof demonstrating that quintic equations could not be solved using ordinary means. In 1824, just prior to Galois's work on group theory, a Norwegian mathematician named Niels Henrik Abel (1802–1829) demonstrated that it was impossible to generate a general formula for the quintic equation using permutations.

Other developments in the study of algebra during the 19th century did not involve forming generalizations of algebraic equations. Research into both number and set theories played an important role in 19th century mathematics, including the study of algebra (see NUMBERS AND NUMBER THEORY). One such development was the application of algebraic principles to the study of sets. This is best illustrated by the work of the English mathematician George Boole (1815–1864). Boole, along with a group of others, studied the mathematical basis of **logic**. Boole was responsible for developing what is called the algebra of logic, which he published in 1834. If there are two sets, X and Y, and we wish to consider the sum of the items in both X and Y, then this is expressed as $X + Y$, which is read as either X and Y, meaning that the items can be in either set X or set Y. If we wish to consider the product of these sets, also called the intersection, then the formula is expressed as XY and read as both set X and set Y. If the intersection of these sets is empty, or the null set, then the product would be $XY = 0$. Boole applied other algebraic principles to the study of algebra, and although his work did not have a tremendous influence on the math of the 19th century, it has become the basis of electronic circuits and computer logic in the 20th century.

The 19th century study of algebra also interacted with the simultaneous development of number theory (see NUMBERS AND NUMBER THEORY). Mathematicians studying both algebra and number theory in the 19th century had the need to express a quantity that was equal to the value of the square root of −1. Since the square root indicates the factor that when multiplied by itself will give the desired product, no rational number can express the square root of −1. The law of signs states that when two negative numbers are multiplied together, the product is a positive number. To remedy this (see NUMBERS AND NUMBER THEORY) mathematicians developed complex numbers, sometimes also called imaginary numbers, and denoted them by the letter i. According to this the square root of −1 equals i, or $i^2 = -1$. While complex numbers eased some aspects of algebraic proofs, it complicated others, one of which was how to plot complex numbers in more than two dimensions. This problem was solved in 1843 when the Irish mathematician William Hamilton (1805–1865) developed quanternions. In the study of geometry mathematicians had determined that it was possible to redefine the **axioms**, or truths, that were used in a proof and in the process developed new ways of examining old problems (see GEOMETRY). Hamilton did the same thing with quanternions. If Hamilton redefined some basic algebraic principles, namely the **commutative law of multiplication**, then he could define complex numbers in a four-dimensional field. This discovery played an important role in later studies of **vectors**. Others, such as the English mathematician Arthur Cayley (1821–1895) developed similar methods of analyzing geometric shapes in dimensions greater than three.

There can be little doubt that the study of algebra underwent a significant change in the 19th century. Through the discoveries of a series of revolutionary mathematicians, the rules of solving algebraic problems and the principles of algebraic thinking were redefined. As was the case with geometry, this process frequently involved abandoning established methods of approaching a problem. Over the course of the century the principles of modern algebraic analysis were established, to be further refined in the 20th century.

Selected Bibliography

Grattan-Guiness, Ivor. *The Norton History of Mathematical Sciences: The Rainbow of Mathematics*. New York: W. W. Norton, 1997.

Katz, Victor J. *A History of Mathematics: An Introduction*, 2nd edition. Reading, MA: Addison-Wesley Educational Publishers, 1998.

Kline, Morris. *Mathematical Thought from Ancient to Modern Times*. New York: Oxford University Press, 1972.

Travers, Bridget (ed.). *World of Scientific Discovery*. Detroit, MI: Gale Research., 1994.

Varadarajan, V. S. (ed.). *Algebra in Ancient and Modern Times*. Providence, RI: American Mathematical Society, 1998.

Anesthetics (1800–1904): The scientific revolution that swept the majority of the scientific disciplines during the 17th and 18th centuries was slow to influence the practice of medicine. Some discoveries, such as the invention of the clinical thermometer by the 17th century Italian physician Sanctorius (1561–1636), enabled medical practitioners to more accurately assess the physical state of their patients, but little could be done in many cases to prolong life. This changed significantly in the 19th century as a result of a number of important discoveries. One of the more important of these was the development of the germ theory of disease in the 1860s by the French scientist Louis Pasteur (1822–1895) and the German scientist Robert Koch (1843–1910). This theory established the role of bacteria as the causing agents of many illnesses (see GERM THEORY), which in turn led to the development of **pasteurization** procedures and antiseptic techniques. By the end of the 19th century, medical professionals had also widely accepted the use of vaccinations to reduce the risk of contracting lethal diseases such as smallpox and cholera (see VACCINATIONS). Furthermore, many areas of medicine were developing specialized instruments to aid in examining specific body parts for aliments (see MEDICAL INSTRUMENTS). In many ways. though, surgical practices of the 19th century differed little from those of earlier times. This was mostly because, to be effective, surgeons had to overcome the problem of pain. While doctors around the world used procedures such as hypnosis, administering locally available herbs, and alcohol intoxication, at the start of the 19th century not a single mechanism was both effective and widely accepted by the medical community. However, this changed in the 19th century with the discovery of chemical compounds called anesthetics that limited the patient's perception of pain. This discovery had a tremendous influence on the modernization of the practice of medicine.

By the end of the 18th century chemists and physicists had developed an interest in describing the properties of gases (see GAS LAWS). In 1792 the English scientist Joseph Priestly (1733–1804) discovered a gas called nitrous oxide (chemical formula: NO). The anesthetic properties of nitrous oxide were first described in 1800 by another English scientist, Humphrey Davy (1778–1829). Davy experimented with using nitrous oxide to put animals to sleep and suggested that the gas would be useful in the practice of medicine. However, Davy also discovered that nitrous oxide produced a light-headed, euphoric feeling in humans and gave it the name laughing gas. Before its use as an anesthetic nitrous oxide was frequently inhaled by Davy and his friends at parties as an intoxicant. During his brief career, the English physician Henry Hill Hickman (1800–1830) experimented with the use of gases such as carbon dioxide to put animals to sleep for surgery, but his work went largely unknown by scientists.

As is frequently the case in medicine, local doctors use innovative procedures to help their patients without reporting their findings to medical journals. However, for such practices to be accepted by the medical community they routinely must be demonstrated before a medical review board or published in a respected journal. Such was the case with the first medical use of anesthesia in Western medicine. There is evidence that the American physician Crawford Long (1815–1878) used a chemical compound called **ether** to anesthetize a patient for surgery in 1842. The anesthetic effects of ether had been suggested much earlier, by the Swiss scientist Paracelsus (1493–1591), who had described the ability of ether, then called sweet vitriol, to introduce sleep, and later (ca. 1818) by the English scientist Michael Faraday (1791–1867), who suggested that it could be used as an anesthetic. Unfortunately Long did not publish his findings until 1849, and then only in a small southern medical journal. Thus, credit for the use of ether as the first anesthetic is frequently given to the American physician William Thomas Green Morton (1819–1868). In 1846 Morton performed a tooth extraction on a patient using ether before an assembly of medical professionals at Massachusetts General Hospital. A subsequent demonstration of a tumor removal using ether as an anesthetic promoted a widespread interest in the procedure, and within a short period of time ether had gained popularity among surgeons.

In 1831 the American chemist Samuel Guthrie (1782–1848) manufactured a chemical compound called chloroform (chemical formula: $CHCl_3$), although the discovery was simultaneously and independently made by the German scientist Justus von Liebig (1803–1873) and by the French scientist Eugene Soubeiran (1793–1858) in Europe. Each noted the possible use of chloroform as an anesthetic, but credit for the first medical use is given to the Scottish obstetrician James Simpson (1811–1870). Simpson was familiar with Morton's use of ether, but he disliked the smell of the compound and thought that it had dangerous side effects. In 1847 Simpson performed the first use of chloroform to reduce the pain of childbirth and immediately came under criticism from religious leaders (mostly men) who regarded the pain of childbirth as a necessary part of the procedure. However, when Queen Victoria requested the use of chloroform anesthetic for the birth of her son, the procedure gained almost instant popularity.

By the end of the 19th century the use of inhaled anesthetics was in widespread use by the medical community. Early in the century the anesthetic had simply been placed into a bowl and the patient inhaled the vapors, but later more elaborate delivery mechanisms were developed in which masks were used to concentrate the anesthetic for the patient. However, these were still general anesthetics. The first compound used to anesthetize a small region during a medical procedure was cocaine. These types of compounds are called local anesthetics because the patient typically remains awake, but often sedated, during the operation. The possible use of cocaine, a compound derived from the South American coca plant, as a painkiller was suggested by the famous Austrian psychoanalyst Sigmund Freud (1856–1939). Credit for the first use of cocaine as a local anesthetic is given to Carl Koller (1857–1844), an Austrian surgeon who was investigating different compounds to be used for eye surgery. Before the use of anesthesia, eye surgery was practically impossible, mostly because it was extremely difficult to keep the patient still during a procedure. In 1884 Koller effectively demonstrated the use of cocaine as an anesthetic during an eye surgery in Germany. Unfortunately, cocaine has a low toxic dose and is highly addictive. Furthermore, supplies of cocaine were rare and expensive to come by in the 19th century. For the next two decades chemists and physicians experimented with the use of various other chemicals as local anesthetics. However, it was not until 1904 that a safe alternative was developed. In that year a compound called procaine, more commonly known as novocaine, was invented. Novocaine is still used by modern physicians, especially in the practice of dentistry.

As the practice of using anesthesia became more common, demands for surgical professionals rose, and by the 20th century all areas of medicine had surgical specialists. Today specialized physicians called anesthesiologists are licensed to deliver the various chemical anesthetics during an operation. Yet, despite the technological advances in the science of reducing pain, the general procedures of inhalation or injection have remained relatively unchanged since the 19th century. Anesthesia is now commonly used for many medical procedures, from general dentistry to childbirth, and represents one of the most significant advances in the history of medicine.

Selected Bibliography

Duffin, Jacalyn. *History of Medicine: A Scandalously Short Introduction.* Toronto, Canada: University of Toronto Press, 1999.

Fradin, Dennis B. *"We have conquered pain": The Discovery of Anesthesia.* New York: Simon and Schuster, 1996.

Margotta, Roberto. *The History of Medicine.* New York: Smithmark Publishers, 1996.

Robinson, Victor. *Victory over Pain: A History of Anesthesia.* New York: Henry Shuman, 1946.

Aspirin (1828–1899): The drug aspirin is one of the most commonly known and used drugs in the world. *Aspirin* is a generic term (see below) for the chemical compound acetylsalicylic acid (see Figure 2), which belongs

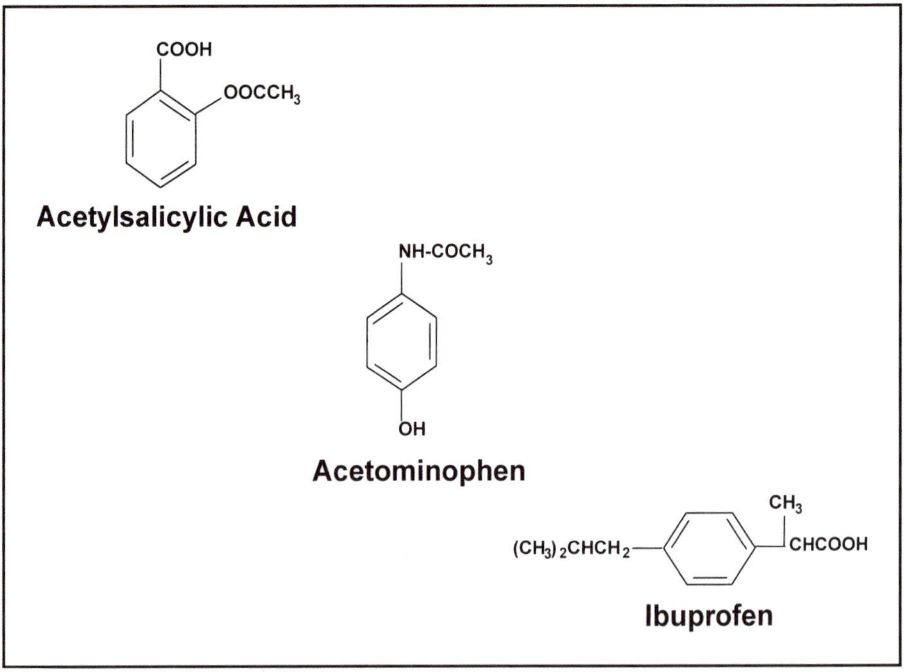

Figure 2. The chemical structures of three common pain-relieving medications. Acetylsalicylic acid is the chemical name for the generic drug aspirin.

to a group of chemicals called the salicylates. Salicylates are found naturally in greatest abundance in the bark of the willow tree, *Salix alba*. Since ancient times medical folklore acknowledged that powdered willow bark helps reduce fever and swelling of wounds. The Assyrian, Egyptian, Greek and native North American cultures record the use of this remedy. However, not until the 19th century was first of these chemicals artificially manufactured, and over the course of the century it was refined to become what is now known as aspirin.

While willow bark remained a folklore medicine in many parts of the world throughout history, the scientific study of the compound did not begin until the late 18th century, when the Reverend Edward Stone of England first called attention to the compound by sending a letter to The Royal Society, a prestigious English scientific organization. The Reverend Stone had been conducting what are now considered to be clinical trials of the effects of powdered willow bark on fever patients. By the early 19th century a number of chemists were working to isolate the active component of the willow bark.

The first success came in 1828 when the German scientist Johann Buchner (1783–1852) first isolated a small amount of salicylic acid. Over the next decade chemists worked to make the extraction process more productive, but it was not uncommon to have to use several pounds of willow bark to obtain a few ounces of the compound. This changed significantly in 1860 when the German chemist Hermann Kolbe (1818–1884) was successful in determining the chemical structure of salicylic acid. Once the chemical structure of the compound had been determined, it was possible to devise a means of synthetically producing large amounts of the chemical on an industrial scale. By 1874 the Germans had a factory operating in Dresden that produced large amounts of the sodium salt form of salicylic acid. The main problem of this form was that it frequently caused stomach problems and had a strong taste.

An alternative form of the drug was invented in 1898 when Felix Hoffmann, (1868–1946) a chemist with the Bayer chemical company of Germany, developed a form of the drug that included an acetyl functional group (see Figure 2). This structure of the drug had the same clinical effects on patients, but greatly reduced the negative side effects. This form of salicylate actually was first invented in 1845 by Charles Gerhardt (1816–1856), who unfortunately did not recognize the potential of the new form. The Bayer Corporation patented the acetylsalicylic acid under the name Aspirin in 1899, and by 1904 they had introduced the pill form of the drug commonly used today. Originally the word *aspirin* was a trademark of the Bayer Corporation. However, following World War I the company lost the rights to that name, and thus there are now many generic forms of the drug.

While the medical effects of aspirin are well documented, the mechanism by which it acts was not determined until the second half of the 20th century. In 1971 the English scientist John Vane (1927–) discovered that aspirin works by inhibiting the action of prostaglandins, chemicals released by damaged tissue. Rather than stop the brain's perception of pain, aspirin stops the manufacture of the prostaglandins at the site of the damaged tissue. However, these same prostaglandins are used by the cells in the lining of the stomach to sense the level of acid and manufacture the protective mucus covering. One of the frequent side effects of aspirin is irritation of the stomach lining caused by prostaglandin inhibition. For his contribution in understanding the role of prostaglandins in the pain response, Vane was awarded the 1982 Nobel Prize in physiology or medicine.

Aspirin is not the only medicine associated with pain relief. Ibuprofen, first made available to the public in 1967, is a compound (see Figure 2) that has the ability to treat fever, pain and inflammation in the same manner as aspirin. This group of compounds, which includes naproxen and ketoprofen, are sometimes called nonsteroidal anti-inflammatory drugs (NSAIDs). NSAIDs do not appear to have the same influence on the circulatory system as aspirin does. A similar medicine is acetaminophen, which also has the fever- and pain-reducing qualities but does not inhibit the inflammatory response. Aspirin is also recognized to both reduce the ability of blood **platelets** to clump together and increase the elastic nature of the blood vessels. For these reasons low doses of aspirin are now recommended as a preventive medicine for cardiovascular disease, with studies suggesting that it increases the survivability of stroke and heart attack patients. However, it is not without its negative side effects. Large doses of aspirin can causes kidney problems, and interactions with aspirin and certain viruses, namely chicken pox, can cause **Reye syndrome**.

Aspirin was not the only drug developed during the 19th century. The advances in chemistry, and specifically organic chemistry (see CHEMISTRY, ORGANIC CHEMISTRY), opened up a new era for chemists and they used their abilities to study the chemical structure and properties of a number of compounds, some of which had applications in medicine. For example, by the end of the century cocaine was being used as an anesthetic (see ANESTHETICS). In addition, barbiturates, first manufactured by the German chemist Adolf von Baeyer (1835–1917) in 1863, became a major component of tranquilizers and sleeping aids in the 20th century. However, of all the drugs developed during the 19th century, aspirin is the most significant. The annual worldwide use of aspirin is not known, but it is estimated that more than 80 billion pills are consumed each year.

Selected Bibliography

Jack, David B. One Hundred Years of Aspirin. *Lancet* 350(9075): 437–439.

Leake, Chauncey D. *An Historical Account of Pharmacology to the 20th Century*. Springfield, IL: Charles C Thomas Publishers, 1975.

Van Dulken, Stephen. *Inventing the 19th Century: 100 Inventions that Shaped the Victorian Age from Aspirin to the Zeppelin*. New York: New York University Press, 2001.

Weissmann, Gerald. Aspirin. *Scientific American* 264(1): 84–90

Asteroids (1801–1898): The discovery of the planet Uranus by the English astronomer William Hershel (1738–1822) in 1781 initiated a series of important changes in the study of planetary astronomy (see PLANETARY ASTRONOMY). Not only was this the first discovery of a planet since the invention of the telescope in the 17th century, but it also marked a renewed interest in detailed examinations of the solar system. But of a greater importance was the linking of mathematics and astronomy. In 1766 Johann Titius (1729–1796) and Johann Bode (1747–1826) suggested that there was a mathematical basis for the spacing of the planets, an idea that had been suggested in various forms since the German astronomer Johannes Kepler's (1571–1630) work on harmonics and planetary motion in the 17th century. The location of Uranus matched exactly the location predicted by the Titius–Bode law. Of far greater significance to 19th century astronomy was that a second gap existed in the solar system between the orbits of Mars and Jupiter and that the Titius–Bode formula indicated that this should be the location of an additional planet. When coupled to the search for Neptune and Vulcan (see PLANETARY ASTRONOMY), the examination of the Mars–Jupiter gap meant that a large number of powerful telescopes were now being dedicated to the search for planets.

At the start of the 19th century a number of astronomers were dedicating their efforts to the discovery of the missing planet. Yet the first major discovery of the century actually occurred by accident. In 1801 the Italian astronomer Giuseppe Piazzi (1746–1826) was updating a series of star charts when he discovered an object that had moved against the backdrop of the night sky. This was not unusual, since the discovery of new comets frequently began with this type of observation. However Piazzi's object did not display any of the characteristics of a comet. After determining that the object was located between Mars and Jupiter, and observing the object for several days to ensure that it was not simply an inaccuracy in his charts, Piazzi announced the discovery of the missing planet. He named the planet after the Sicilian goddess Ceres. Unfortunately, the new planet was very small (diameter, 640 miles), and Piazzi was able to make only a few measurements of its orbit before he lost it against the night sky. Luckily for Piazzi the German mathematician Carl Friedrich Gauss (see ALGEBRA) had developed a method of plotting orbits with a minimum of points. Using Gauss's formula it was possible to calculate the approximate orbit of Ceres, and within a year of its disappearance the "planet" was rediscovered.

Although small by planetary standards, Ceres was in the correct location according to the Titius–Bode law. However, within a short time of its discovery astronomers began to realize that something was wrong. In 1802 the German astronomer Heinrich Olbers (1758–1840) discovered a second planet in close proximity to Ceres. This object was given the name Pallus. Astronomers were puzzled by the closeness of Ceres and Pallus, but a greater problem was the fact that the **inclination** of Pallus's orbit was different from that of Ceres. After studying the size and orbits of these new "planets," Herschel suggested that they should not be considered as true planets but, rather, as small starlike objects, which he called *asteroids* after the Greek word for star. Others used the term *planetoid* to represent that these objects behaved like minor planets. By 1807 two additional asteroids, named Vesta and Juno, were discovered in the same area. Astronomers were beginning to recognize that instead of a planet between Mars and Jupiter there were a large number of these asteroids. This area, consisting of several thousand identified asteroids, is now commonly called the asteroid belt. Some astronomers estimate that as many as 10,000 asteroids greater than a mile in diameter may occupy this region.

While asteroids are common in the solar system, modern astronomy now recognizes that they are not distributed evenly between the planets. This was first suggested in 1866 when the American astronomer Daniel Kirkwood (1814–1895) discovered that distinct regions in the solar system lacked asteroids. He suggested that these regions were to the result of the gravitational interference of Jupiter, which produces resonance zones that influence smaller bodies such as asteroids. In general, Jupiter's intense gravitational field either traps asteroids or pushes them into more distant orbits outside the planet's zone of gravitational influence. Several of these resonance zones have been identified, and collectively they called the Kirkwood Gaps.

By the late 1800s the use of photography had already provided a technological advance for the study of astronomy (see ASTRONOMY, PHOTOGRAPHY). First, it allowed astronomers to take pictures of the night sky, thus providing astronomers with the ability to study a section of the sky for a longer period of time. Furthermore, the use of photographic plates greatly simplified the task of discovering new asteroids. Whereas in the past astronomers were forced to make a series of detailed nightly observations to detect new asteroids, it was now possible to detect them using photographic techniques. To do this, a telescope with a camera attached is positioned so that it rotates at the same rate as the earth. This type of arrangement is called an **equatorial mount**. The shutter of the camera is then left open for a constant exposure. Since the telescope and stars appear to moving at the same rate, the stars appear as single dots of light. But any object that is in motion, such as a comet or asteroid, appears as a white line against the dark sky. Once detected, more detailed observations may be made to determine the size, location and orbit of the object. This procedure was first used in 1891 by the German astronomer Max Wolf (1863–1932) in his discovery of the asteroid named Brucia. Wolf identified more than 240 asteroids using this procedure, and it remains the primary mechanism for detecting asteroids in modern astronomy.

For the majority of the 19th century the study of asteroids had been confined to those that were located within the asteroid belt between Mars and Jupiter. However, in 1898 Gustav Witt, a German astronomer, discovered Eros. The orbit of Eros takes it from the asteroid belt to inside the orbit of Mars. In fact, Eros approaches within 14 million miles of Earth and represents what is called a "near-Earth" asteroid. Recently (2001) Eros was the focus of a NASA project called the Near Earth Asteroid Rendezvous (NEAR). As part of this project a probe called NEAR-Shoemaker was placed in orbit around Eros to map the asteroid and eventually crash-landed on the asteroid to determine its composition. Modern astronomers now group asteroids on the basis of their orbits. In recent years the studies of the near-Earth asteroids have taken on a special significance as astronomers, biologists, geologists and paleontologists uncover evidence that the evolutionary history of our planet has been shaped by collisions with these objects in the past. Called planet killers because of their size (>10 km in diameter), these objects differ from the usual meteorites that strike the planet daily. These studies employ many of the same photographic and observational techniques that were adopted in the 19th century when the study of asteroids first became popular among astronomers.

Selected Bibliography

Kowal, Charles T. *Asteroids: Their nature and utilization*, 2nd ed. New York: John Wiley & Sons, 1996.

Littmann, Mark. *Planets Beyond: Discovering the Outer Solar System*. New York: John Wiley & Sons, 1988.

Peebles, Curtis. *Asteroids: A History*. Washington, DC: Smithsonian Institution Press, 2000.

Astronomy (1838–ca. 1887): Astronomy is an ancient science, with historical evidence indicating that it has been studied by all of the world's major cultures. For most of recorded history astronomers have gathered their information using naked-eye observations of the night sky. While this was useful for the prediction of the seasons and the making of early calendars, it frequently did not provide enough information to develop accurate theories on the nature of the universe. The science of astronomy has frequently been confused with the belief in astrology. Astrology is the idea that the physical positions of the planets and moon influence the daily lives of humans. Ancient astronomers were most likely also astrologers, since the position of the planets, sun and moon did indicate the occurrence of some important events such as harvesting. The difference is that astrology is based on beliefs and has never been proven using any scientific method. However, throughout the ages the science of astronomy has gone through a series of revolutionary periods, each one of which has served to foster a deeper understanding of the complexity of the universe. Perhaps the greatest change in the study of astronomy was the result of the 17th century invention of the telescope. Although early telescopes were primitive, and frequently had less magnification power than modern binoculars, they opened up the night sky for observation. With these early telescopes it was possible to discover new moons and, eventually, planets and asteroids (see ASTEROIDS, PLANETARY ASTRONOMY). Over the next two centuries there were consistent improvements to the optical system of telescopes. In the 19th century the telescope would once again play an important role in astronomy.

Since early times astronomers had been attempting to estimate the size of the universe. In the 17th century a procedure called parallax analysis became popular for determining the dimensions of the solar system. In parallax analysis an object is viewed from two equidistant vantage points. The amount that the object appears to move is called the parallax, and it may be used to estimate distance by geometric methods (see GEOMETRY). This can be illustrated by first holding a pencil at arm's length and then alternately looking at it through the left and then the right eye. The pencil appears to jump in relation to the background. The further the pencil is away from your eyes, the less that it appears to move. In the 17th century this procedure was used to estimate the distance to Mars and other planets. However, the stars in the night did not display a parallax shift, and therefore they were justifiably considered to be located very far from earth. 19th century improvements in telescope lens design, most to the result of the efforts of Joseph von Fraunhofer (1787–1826), helped to remove some of the distortion in a star's image when viewed through the telescope. This made it possible to detect small changes in the position of a star when viewed from different locations on Earth. Another of Fraunhofer's instruments, the heliometer, made it possible to accurately measure the apparent distances between stars. The heliometer was an adaptation of a 17th century instrument called a **micrometer**, which had been used successfully to determine angular distances between stars. The heliometer was actually a hybrid of a telescope and a micrometer and, like Fraunhofer's lenses, his heliometer was a high-quality instrument and thus played an important role in the estimating of astronomical distances.

Using one of Fraunhofer's lenses and heliometers, in 1838 the German astronomer Friedrich Bessel (1784–1846) was able to calculate the distance from a star called 61 Cygni (the 61st identified star of the constellation Cygni) to two other nearby stars. Others, such as the German scientist Wilhelm Struve (1793–1864), had previously used the same approach with the star Vega, but the resulting estimates were not considered to be very accurate. Since the Earth moves in its orbit, Bessel was able to take monthly measurements of 61 Cygni's position from slightly different vantage points. In doing so he calculated that 61 Cygni was more than 64 trillion miles away, or approximately eleven **light-years**. Later determinations of the distance to Vega (27 light-years), Sirius (9 light-years) and Alpha Centauri (4.5 light years) further demonstrated the power of this method. Astronomers of the 19th century considered the brightest stars to be the closest to Earth, and thus if 61 Cygni, Vega and Alpha Centauri, some of the brightest stars in the night sky, were this far away, then the dimmer stars must be even more distant. This suggested that the universe was vastly larger than expected. Since the time of the Copernicus (1473–1543) and Galileo (1564–1642) the view of the Earth as the most important object in the heavens had been changing. Bessel's discovery further reduced the significance of our planet in the universe.

Other instruments designed in the 19th century played important parts in the advancement of astronomy. Studies of the electromagnetic spectrum early in the century had indicated that light consisted of a wide variety of wavelengths, some of which exceeded the detection abilities of the human eye (see ELECTROMAGNETIC SPECTRUM). In 1814 Fraunhofer discovered that the electromagnetic spectrum of sunlight was not continuous, but rather consisted of a series of distinct bands. Around 1860 an instrument called a spectroscope was invented to facilitate the analysis of these bands (see SPECTROSCOPE). The spectroscope enabled astronomers to determine the chemical composition of stars, but it also played an important role in the discovery that the stars themselves were in motion.

Around 1842 Christian Doppler (1803–1853), an Austrian physicist, conducted some important experiments with sound. Doppler noticed that the sound of a locomotive train appeared to be higher in **frequency** as a train approached and lower in frequency after the train had passed. This was due to a compression of the sound waves as the train approached and a subsequent lengthening of the waves after it had passed. Doppler suggested that since light and sound waves were similar, then this approach could be used to measure the motion of stars. Doppler suggested that the color of the star's light should appear to be redder if the star was moving away and would appear to be more blue if the star were approaching. While the color of the star's light is difficult to determine directly, the invention of the spectroscope and Fraunhofer's discovery of the banding patterns of starlight (see SPECTROSCOPE) provided a useful measurement tool. By examining the shift in the dark band areas of the light, it was possible to determine that most stars are moving away from us, a phenomenon called redshift. This was an important discovery in that it allowed for later calculations on the expansion rate of the universe, which in turn helped to estimate its overall age. The study of the Doppler effect in stars also led to the American astronomer Antonia Maury's (1866–1952) discovery in 1887 that in some cases what appeared to be single stars were actually binary stars rotating around a common point in space.

Astronomers now had the ability to determine the distance to a star and the motion of the star. Using this information it was evident that a star's brightness was determined by the nature of the star, called its luminosity, and its distance from Earth. The apparent brightness of a star had been measured as far back as the ancient Greeks. The great Greek astronomer Hipparchus (ca. 190–ca. 120 BCE) had identified the brightest stars as belonging to the first magnitude, the next brightest group as being the second magnitude, and so forth. In other words, as the magnitude increased, the brightness of the star decreased. In 1856 the English scientist Norman Pogson (1829–1891) established a mathematical basis for magnitude. Pogson observed that first-magnitude stars were 100 times brighter than sixth magnitude stars. By using an instrument called a **photometer** attached to a telescope and a series of first-magnitude stars as reference points, it was now possible to accurately determine the stars' apparent magnitudes as viewed from Earth.

By the end of the century astronomers were developing a catalog system for stars based on magnitude. However, not all stars are the same—some have much higher surface temperatures than others. Using photographic plates (see PHOTOGRAPHY) it was possible to further classify stars based on the spectrum of light being emitted and the temperature of the star. The development of photographic techniques played an important role in many areas of astronomical study (see ASTEROIDS, PLANETARY ASTRONOMY, SUN), and by the end of the century a significant number of the larger observatories had been equipped to produce permanent photographic records of the night sky. This historical database has served as an invaluable record for the science of astronomy, and modern observatories frequently conduct more photographic studies than direct-eye studies.

Modern astronomy can be said to have begun with these 19th century discoveries. By the end of the 19th century astronomers had the necessary tools to examine the universe in substantial detail and the ability to quantify their observations. What they discovered was that the universe is immensely more complex than had been originally thought. Instead of answering questions regarding our universe, these 19th century observations set the stage for a series of discoveries that persist to the present day. Modern astronomers still utilize these methods in their study of the night sky, a tribute to the importance of 19th century astronomy.

Selected Bibliography

Hoskin, Michael. *The Concise History of Astronomy*. New York: Cambridge University Press, 1999.

Motz, Lloyd, and Jefferson Hane Weaver. *The Story of Astronomy*. New York: Plenum Press, 1995.

Spangenburg, Ray, and Dianne Moser. *The History of Science in the Nineteenth Century*. New York: Facts of File, 1994.

Trefil, James, "Puzzling out Parallax." *Astronomy* 26, no. 9 (1998): 46–51.

Atomic Theory (1803–1810): The idea that all matter consists of individual atoms first originated long before the 19th century. The ancient Greek philosopher Democritus (ca. 460 B.C.E.–ca. 370 B.C.E.) suggested that all matter was composed of small indivisible particles, which he called atoms. However, little experimental evidence supported this claim, and the study of atomic theory remained dormant for centuries. During the scientific revolution of the late 16th and 17th centuries there developed an interest in explaining the natural world in scientific terms. The English scientist Robert Boyle's (1627–1691) investigations into the properties of air pressure suggested that air consisted of small, individual particles surrounded by a void that could be compressed. This theory of matter, also called the corpuscular theory, was expanded on later in the 17th century by the English scientist Isaac Newton (1642–1727). Newton believed that these individual particles interacted with one another through physical forces of nature. Furthermore, he contended that the strength of these forces decreased with the distance between the atoms according to the **inverse-square law**. To Newton, the same principles that held the planets in their orbits also applied to movement of atomic particles.

By the end of the 18th century there was considerable debate as to whether the ratio of elements in a compound was dependent on the method by which the compound was prepared. Much of this debate was due to inaccurate determinations of **atomic weight** (see ATOMIC WEIGHT). In 1794 the French chemist Joseph-Louis Proust (1754–1826) developed the law of definite proportions. Proust proposed that the elemental components of a compound did not vary. While ideas such as this had been proposed by chemists earlier in the 18th century, Proust was the first to determine this law experimentally. In detailed experiments with copper compounds, specifically copper carbonate, Proust determined that there was a fixed ratio of copper, oxygen and carbon present (5:4:1) and that this ratio remained stable regardless of how the copper carbonate was prepared. The German chemist Jeremias Richter (1762–1807) proposed that these ratios demonstrated that chemistry could be approached as a form of applied mathematics, which he called **stoichiometry** (see CHEMISTRY). Using Proust's law and the principles of stoichiometry, it was possible to distinguish between compounds and their fixed ratios of elements and mixtures whose contents varied dependent on preparation.

These advances at the turn of the century had a strong influence on the work of the English chemist John Dalton (1766–1844). Dalton is responsible for synthesizing the Greek view of the atom with the work of Proust and Richter. While Dalton is considered to be the father of modern chemistry, he in fact started his career with an interest in meteorology (see METEOROLOGY). As a meteorologist, Dalton had an intense interest in the structure of the atmosphere. Specifically he was interested in learning why the gases of the atmosphere did not settle out so that the heavier compounds were closer to the ground. At the time many reasoned this was because the gases of the atmosphere formed a type of homogenous mixture. However, Dalton believed that the gases were actually individual particles and that the atmosphere was held together by the physical, not chemical, attraction of these particles. Dalton performed a number of experiments with gases during the late 18th and early 19th centuries. He performed studies on the partial pressures of gases and also investigated the solubility of gases in water. It was these studies that led to the development of what is called the modern atomic theory.

First, Dalton recognized, as had Democritus and Newton before him, that atoms were small, indivisible particles of matter. However, unlike many of the chemists before him, Dalton first presented the idea that these individual atoms have a specific weight and combine in predictable ratios. Dalton began his study with the chemical interactions of nitrogen and oxygen in the atmosphere. In his studies of nitrogen dioxide (NO_2), nitric oxide (NO) and elemental nitrogen (N) Dalton noticed that these compounds form in specific ratios of the elements. Applying Proust's law of definite proportions and Richter's principles of stoichiometry, Dalton was able to develop his law of multiple proportions around 1803. This law states that the chemical formula of a compound is always a whole number integer ratio of the elements. No partial atoms are involved in the formation of a chemical compound. This law was applied to gases in 1808 by the French chemist Joseph-Louis Gay-Lussac (1778–1850) when he developed his law of combining volumes (see GAS LAWS).

Dalton then studied carbon dioxide (CO_2) and carbon monoxide (CO). He noted that, as was the case with nitric oxide and nitrogen dioxide, the simplest compound that can be formed from two elements is always a binary compound, that is, it contains one of each element. Dalton called this the Rule of Simplicity. This rule uses the law of definite proportions to predict the combinations of elements in a compound. For example, if two elements form two different types of compounds, the first will be binary (for example, NO) and the second will be tertiary (NO_2 or N_2O). While this rule applies to many oxides, it does not apply to all compounds. For example, Dalton mistakenly believed that water would consist of one atom each of hydrogen and oxygen (HO) instead of its correct formula, H_2O.

As individual components of matter, atoms of an element should have a unique weight. Using his law of definite proportions, Dalton believed that he could determine the atomic weights of the elements involved. For example, in the case of water, the 18th century French chemist Antoine Lavoisier (1743–1794) had determined that water was 85% oxygen by mass, and 15% oxygen. Using his idea that water was a binary compound (HO), and using hydrogen as a reference point, Dalton calculated the atomic weight of oxygen to be 5.66 (85 divided by 15). Although this is incorrect, it was an important advance in the study of atomic weights (see ATOMIC WEIGHT).

Each of these advances by Dalton was possible only when elements were composed of distinct atoms. Dalton's work serves as the background for modern atomic theory, which states that (1) all matter is made of atoms, (2) atoms of a given element have a unique mass, and (3) atoms interact in chemical reactions as whole-number ratios. Presented between 1808 and 1810 as the *New System of Chemical Philosophy*, Dalton's ideas on atomic theory would have a tremendous impact on the course of chemical science. Later in the 19th century other chemists would apply Dalton's theories to the study of gases. As mentioned, Gay-Lussac applied Dalton's atomic theory to the study of how gases combine (see GAS LAWS). The Italian chemist Amedeo Avogadro (1776–1856) further expanded on Dalton's work when he discovered in 1811 that a simple relationship exists between the volume of a gas and the molecules that form the gas. This led to the establishment of the

Avogadro constant (or number) and the concept of molecular volumes and weights in the study of gases (see GAS LAWS).

Dalton's atomic theory was not entirely without error. He incorrectly assumed that all atoms of a given element have the same weight. It is now recognized that there is some variation in the atomic weights of individual atoms of a given element. These are called isotopes, and some of them have found important applications as radioactive markers in 20th century science. While Dalton's theory led to the incorrect calculation of atomic weights for elements such as oxygen, these were relatively minor problems that did not directly influence the science of the times. During the early 19th century, chemistry was still a young science, having been established only two centuries earlier. The philosophical advances in the area of atomic theory in the 19th century would modernize the study of chemistry and set the stage for a large number of important chemical and physical discoveries of the late 19th and early 20th centuries.

Selected Bibliography

Cobb, Cathy, and Harold Goldwhite. *Creations of Fire: Chemistry's Lively History from Alchemy to the Atomic Age.* New York: Plenum Press, 1995.

Nye, Mary Jo. *Before Big Science: The Pursuit of Modern Chemistry and Physics 1800–1940.* New York: Twayne Publishers, 1996.

Rocke, Alan J. *Chemical Atomism in the Nineteenth Century: From Dalton to Cannizzaro.* Columbus, OH: Ohio State University Press, 1984.

Atomic Weight (1803–1897): It is well recognized in modern chemistry that the atom is the smallest particle of matter and that the formation of chemical compounds is the result of the interaction of individual atoms. While the concept of an atom dates from the time of the ancient Greeks, the development of the modern atomic theory and the revolution in chemistry that it created are primarily the result of the work of a handful of 19th century chemists (see ATOMIC THEORY). By the mid-19th century the atomic nature of matter was widely accepted in the scientific world. However, the structure of an atom remained relatively unknown. In the 19th century the discovery of the electron and two other developments, the concepts of atomic weight and **valence** (see VALENCY), began a scientific investigation of the atom that would eventually result in the beginnings of the Atomic Age in the 20th century.

The study of atomic weight in the 19th century began around 1803 with the work of the English chemist John Dalton. Dalton's study of atmospheric gases (see GAS LAWS) led him to determine that the elements of a compound exist as whole-number ratios. This was called the law of definite proportions, and it played a key role in the formation of the atomic theory (see ATOMIC THEORY). Using this law, Dalton returned to some early calculations of the 18th century by the French chemist Antoine Lavoisier. Lavoisier had examined the proportions of elements in many common compounds. For example, in the case of water Lavoisier had determined that it was 85% oxygen and 15% hydrogen by volume. Since no mechanism yet existed to directly measure the weight of an atom, the process had to be conducted mathematically. Dalton believed that water consisted of a single molecule each of hydrogen and oxygen. Using hydrogen as a reference, he calculated the atomic weight of oxygen to be 5.66 (85 divided by 15). He used this type of calculation as the basis for one of the first tables of atomic weight.

However, not all scientists of the early 19th century accepted the concept of an atom as an indivisible particle and thus were reluctant to adopt the concept of an atom having a specific weight. Although these scientists recognized that elements existed in compounds according to fixed ratios (see ATOMIC THEORY), they utilized different standards for determining the weights of compounds and the elements within them. These were frequently called equivalent weights, and although they functioned along much the same principle as an atomic weight, they frequently used other elements, namely oxygen, as the reference point. It was not until 1815 that the English chemist William Prout (1785–1850) suggested that hydrogen was the fundamental building block of all matter and based all atomic weights on multiples of hydrogen. Hydrogen was determined to have a weight of one (1) as this reference point. However, this idea was also not widely accepted, and as a result chemists debated the practice of using atomic or equivalent weights until the mid-19th century. The result was the construction of several different tables describing atomic weights, such as Dalton's list of atomic weights (1810) and William Wollaston's (1766–1828) equivalents table (1814). Wollaston's table is an excellent example of the use of oxygen as a reference molecule.

The most notable of these early atomic weight tables was constructed by the Swedish chemist Jöns Berzelius (1779–1848) in 1818. Berzelius was an influential chemist who made a number of significant contributions in the identification of elements and the development of general chemistry (see CHEMISTRY, ELEMENTS). By 1818 a number of scientists studying the properties of gases had provided information to support Dalton's concept of an atom (see GAS LAWS). Most important of these was the work of Amedeo Avogadro. Avogadro corrected Dalton's idea that water had the molecular formula HO, and instead demonstrated that water consisted of two hydrogen atoms. This became the foundation of what is called Avogadro's Hypothesis, which states that if the **density** of one gas is twice that of another, then the first gas must contain twice the number of particles as the second gas. This was an important first step in the measuring of chemical reactions, also called qualitative chemistry (see CHEMISTRY). While Avogadro's work should have corrected Dalton's errors on atomic weights, the correction was widely ignored until later in the century. Unfortunately Berzelius did not recognize Avogadro's discovery until 1826, and many of the initial atomic weights he calculated assumed that the compounds were binary (one atom of each element). However, he did conduct several thousand atomic weight experiments and designed one of the most detailed tables of atomic weights of the early 19th century. Many of his values closely approximated modern values, a significant accomplishment given the instruments of the time. Berzelius also developed the term *molecular weight* to signify the weight of a single molecule of a compound, although the term did not receive much attention until the 20th century.

The debate over the nature of atomic weights continued throughout the century, with scientists divided over the use of equivalent or atomic weight. However, as the century progressed additional experimental evidence was accumulating to support the concept of atomic weight. In 1819 the French scientists Alexis Petit (1791–1820) and Pierre Dulong (1785–1838) published a paper demonstrating a relationship between the **specific heat** of a substance and its atomic weight. The relationship, later called Dulong–Petit law, allows for the calculation of the atomic weight if the specific heat of a substance is known. Studies of the volumes of gases by Avogadro and the relationships of the density of a vapor to atomic weight by the French scientist Jean Baptiste Dumas (1800–1884) in 1826 provided further evidence of an atomic weight, although Dumas was himself not a supporter of the atomic theory. By the mid-19th century influential chemists such as Charles Gerhardt and Stanislao Cannizzaro (1826–1910) had developed atomic weight tables that were widely accepted by chemists.

Atomic weights were put to practical use in the 19th century in the identification and classification of elements. In his construction of a periodic tables of the elements (see ELEMENTS), the Russian chemist Dmitri Mendeleev (1834–1890) predicted the existence of certain missing elements based on gaps in atomic weights when he organized his elements. Mendeleev was so sure that the elements were arranged by atomic weight that he frequently revised weights that had been determined experimentally to match the predicted weight indicated by his table. While not a preferred method of conducting science, it often had a beneficial effect by increasing the accuracy of Mendeleev's table. In 1894 the discovery of the inert gas argon was aided by the knowledge of atomic weight. The English scientists Lord Rayleigh

(1842–1919) and William Ramsey (1852–1916) had noticed that atmospheric nitrogen possessed a higher atomic weight than could be explained theoretically. They designed an experiment using a new procedure called **spectral analysis** (see LIGHT) and in the process discovered the presence of argon.

The study of atomic weights in the 19th century was important because it allowed scientists to begin the exploration of the atom. The debate over the use of atomic or equivalent weights promoted a significant amount of scientific research into the nature of compounds and chemistry in general. Although the identification of isotopes, atoms of an element with different atomic weights, in 1913 would present some challenges to the idea that all atoms of an element possess the same atomic weight, it would lead to the beginning of the Atomic or Nuclear Age in the second half of the 20th century. The groundbreaking work of these 19th century chemists provided the stimulus for much of 20th century nuclear chemistry.

Selected Bibliography

Cobb, Cathy, and Harold Goldwhite. *Creations of Fire: Chemistry's Lively History from Alchemy to the Atomic Age.* New York: Plenum Press, 1995.

Nye, Mary Jo. *Before Big Science: The Pursuit of Modern Chemistry and Physics 1800–1940.* New York: Twayne Publishers, 1996.

Taton, Rene (ed.). *History of Science: Science in the Nineteenth Century.* New York Basic Books, 1965.

B

Biomolecules (1806–1908): Biomolecules are organic compounds that are used in the building of a cell, the fundamental unit of life. Study of these molecules in the 19th century began initially with the founding of the science of organic chemistry and the subsequent development of an understanding of how carbon molecules are structured (see ORGANIC CHEMISTRY). This linking of organic chemistry with biological processes and metabolic pathways is frequently called *biochemistry*. Although biochemistry as a science was in its infancy in the 19th century, as mechanisms to investigate biochemical pathways were just developing, some early work was being done on the process of fermentation (see FERMENTATION). However, during this time a significant amount of information was being accumulated on the major classes of biomolecules. This data would have a strong influence on the study of biochemistry in the following centuries.

The four general classes of biomolecules are proteins, carbohydrates, fats and lipids, and nucleic acids. In modern biochemistry proteins are recognized to be the working molecules of the cell, with almost all metabolic processes involving at least some level of protein action. In the latter part of the 18th century and into the early 19th century scientists had begun to isolate a number of compounds from living tissues that, when heated, did not melt but rather assumed a solid form. This class of molecules was initially given the name *albuminous*. Included in this class were proteins such as albumin from egg whites, globulins from blood, and milk protein called casein. However, the term *protein* was not yet applied to these compounds, and their role in living tissues was not understood. At around the same time, chemists were isolating another group of compounds that are now called the amino acids. Two of these, asparagine (1806) and cystine (1810), had been discovered in the early part of the century as scientists were studying the chemical composition of living cells. While amino acids are now known to be the building blocks of proteins, in the early 19th century the connection between amino acids and proteins had not been established. The first amino acid that was clearly identified as being associated with the albuminoids was glycine (Figure 3). In 1820 Henri Branconnot (1781–1855) isolated a sweet compound from animal tissue that he thought to be a form of carbohydrate. On further investigation he discovered that the compound contained nitrogen, an element that is absent in sugars. At the time the only molecules that were known to contain

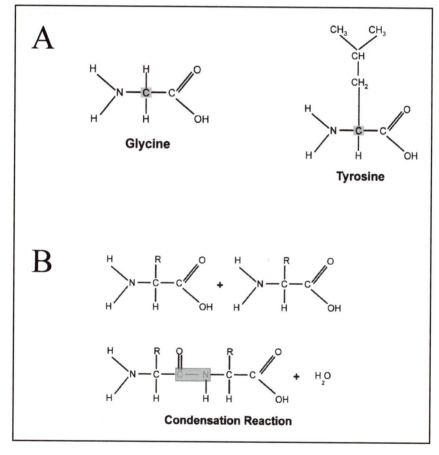

Figure 3. The structure of amino acids and proteins. The top two diagrams (*a*) represent the structure of two common amino acids studied during the 19th century, glycine and tyrosine. The bottom diagram (*b*) represents a condensation reaction between two amino acids, forming a peptide bond.

nitrogen were the albuminoids. The discovery of other amino acids, such as leucine and tyrosine, followed. While these newly discovered amino acids obviously had some function in living cells, their role in the formation of a functional protein was unclear.

The relationship between amino acids and proteins began to take shape around 1839 with the work of the Dutch chemist Johannes Mulder (1802–1880). Mulder was investigating the albuminoids when he discovered what he thought to be a common **empirical** formula, $C_{40}H_{62}O_{12}N_{10}$. Mulder believed this to be the backbone of all albuminoids, to which other elements such as sulfur and phosphorous could be attached. Since these molecules had the ability to deliver all of the major elements (carbon, hydrogen, nitrogen, oxygen, sulfur and phosphorous) needed for living cells, Mulder and the Swedish chemist Jöns Berzelius named the compounds *proteins*, from the Greek word for first importance. By this time other chemists had begun to work out the structural formula for the amino acids. In their investigations it was noted that all amino acids consist of the same core structure, with variations attached to the central, or alpha, carbon (see Figure 3). Furthermore, advances in chemical procedures toward the end of the century had made it possible to isolate amino acids directly from proteins; what was missing was the mechanism by which proteins were formed. This was determined in 1901 by the German chemist Emil Fischer (1852–1919). Fischer demonstrated that amino acids are formed by a **condensation reaction** between the amino end of one amino acid and the carboxylic acid end of a second amino acid (see Figure 3). By 1908, Fischer had used this principle to artificially construct a protein consisting of eighteen amino acids. While there remained considerable debate in the 20th century on the theoretical limit to the size of proteins, there was no doubt as to their structure and relationship to amino acids.

Proteins were not the only biomolecules studied in the 19th century. The class of biomolecules called the carbohydrates was recognized well before the 19th century as being an important energy source for living organisms. Modern biochemists classify carbohydrates as having a $1:2:1$ ratio of carbon, hydrogen and oxygen. In the 19th century Fischer was studying the structure of glucose, a simple sugar with the empirical formula $C_6H_{12}O_6$. The German chemist Heinrich Kiliani (1855–1945) had previously determined (1886) the structure of glucose. However, other sugars were known to have the same chemical formula as glucose, such as mannose and fructose. Fischer developed chemical methods to study the structure of these compounds. What he discovered was that although these molecules have similar empirical formulas, they possess different structures (see Figure 4). These types of compounds are called *isomers*, and their discovery was a significant breakthrough

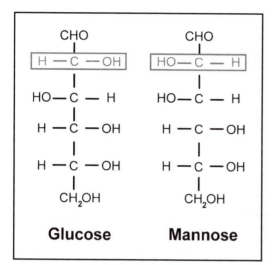

Figure 4. An example of two isomers. These are the monosaccharides glucose and mannose. Notice that they have the same molecular formula $(C_6H_{12}O_6)$ but different structures.

in biochemistry. For his work, Fischer received the 1902 Nobel Prize in chemistry.

Another of the energy molecules, the fats, also received attention from the scientific community in the 19th century, although from a slightly different perspective than the other biomolecules. As with the other biomolecules, scientists in the early part of the century had been trying to determine the structure of fats and lipids. Around 1825 the French chemist Michel Chevreul (1786–1889) determined that the fats consisted of two components, glycerol and fatty acids. It is now recognized that these components have very different chemical characteristics, with glycerol behaving in a similar manner to a sugar, while the fatty acids are long, **hydrophobic** structures. In his studies Chevreul isolated several different forms of fatty acids, including the common oleic, palmitic and stearic fatty acids. However, although this was an important discovery for chemistry, it had a much greater impact on the culture of the time. Until this time candles were made primarily from an animal compound called tallow. The discovery of the fatty acids allowed the manufacture of improved candles that burned cleaner and brighter and smelled better, thus quickly replacing the older versions.

The last major class of biomolecules, the nucleic acids, was initially not a major focus of 19th century biochemists. However, their discovery had a tremendous impact on the direction of science in the 20th and 21st centuries. In 1869 the Swiss chemist Johann Miescher (1844–1895) detected a unique group of chemicals in the nucleus of the cell. Unlike the other classes of biomolecules, these chemicals contained both phosphorous and nitrogen. These same compounds were also detected in the nuclei of other cells. This class of compounds was initially called nuclein. In 1885 the German biochemist Albrecht Kossel (1853–1927) determined that there were actually two different types of nuclein, one that was made of protein and a second that was not a protein. The chemicals that make up this second type are called the nucleic acids. There are five distinct forms of nucleic acids, which were classified into two major groups. The purine nucleic acids consisted of the compounds adenine and guanine, and the pyrimidines consisted of cytosine, thymine and uracil. Kossel received a Nobel Prize in physiology or medicine in 1910 based partly on this work. Little additional work was done on nucleic acids in the 19th century. Although nucleic acids are found in the nucleus of the cell, their connection with heredity did not occur until a series of discoveries later in the 20th century (see INHERITANCE). Scientists today recognize that nucleic acids have several roles in the cell, including heredity and metabolism. Deoxyribonucleic acid, or DNA, belongs to this class of biomolecules, although it was not until 1953 that James Watson (1928–) and Francis Crick (1916–) discovered the helical structure of this important molecule.

These discoveries of the 19th century played an important role in the development of 20th century biochemistry. Each of these achievements enabled scientists to develop an understanding of the structure of biomolecules, an important first step toward the ability to manipulate the chemicals in a laboratory. This understanding resulted in significant 20th century advances in the study of medicine, agriculture and genetics. Many modern drugs and food additives or substitutes were engineered utilizing the structural information first developed in the 19th century. Modern society is completely reliant on chemicals engineered from these biochemicals, a tribute to the first generation of biochemists.

Selected Bibliography

Asimov, Isaac. *Asimov's New Guide to Science*. New York: Basic Books, 1984.

Fruton, Joseph H. *Molecules and Life: Historical Essays on the Interplay of Chemistry and Biology*. New York: John Wiley & Sons, 1972.

Nye, Mary Jo. *Before Big Science: The Pursuit of Modern Chemistry and Physics 1800–1940*. New York: Twayne Publishers, 1996.

Serafini, Anthony. *The Epic History of Biology*. New York: Plenum Press, 1993.

C

Calculators (ca. 1820–1892): The first form of calculating machine was most likely the abacus. The modern version of the abacus, with ceramic beads attached to strings, originated with the Chinese and dates back to the 11th century, although versions were constructed by the Egyptians centuries earlier. In the 1700s the English mathematician William Gunter (1581–1626) invented a mechanical slide rule that assisted with the manipulation of logarithms. While both the abacus and slide rule assisted with calculations, most historians do not consider them to be calculators since both instruments required the operator to move parts to conduct the calculation. The first machine that actually performed calculations mechanically was built by the German mathematician William Schickard (1592–1635) around 1623. However, credit for the design of the first calculating machines to be produced commercially is often bestowed on the French scientist Blaise Pascal (1623–1662) who invented a machine that performed both addition and subtraction around 1642. In 1674 the German scientist Gottfried von Leibniz (1646–1717) invented the first machine that would mechanically perform both division and multiplication operations. The basis design of these two machines remained the standard for the small number of calculating machines in use through the 18th century. However, in the 19th century a few inventors made some important changes to the design of the calculator that enhanced both its calculating power and attractiveness to mathematicians.

The precursor of the 20th century computer, which has the capability of performing billions of calculations per second, actually originated in the 19th century. Its inventor, Charles Babbage (1791–1871), supposedly developed an interest in building a calculating machine after discovering numerous errors in astronomical charts used by the maritime industry. Starting in 1820 Babbage began to work on what he called a difference engine, a machine capable of performing repetitive subtraction and addition operations without error. The challenge was that this machine was being designed well before the invention of electronic circuits, and thus all calculations had to be performed mechanically. Furthermore, to construct a device capable of detailed calculations would require the use of precision-designed materials. Babbage's initial designs called for a machine that was massive, more than 6 feet long and wide. The device had a series of wheels connected to a single axis. In the first axis the operator entered the desired number. The second axis contained the value to be subtracted (or added). Using a system of levers and cams, this value was subtracted or added sequentially in the later axis. Thus the fourth axis of wheels would indicate the result of subtracting the quantity twice from the entered value, and so on. Babbage envisioned that the machine would be connected to a printing device to provide an output of the operations. However, because of funding problems and conflicts with his investors, the machine was never constructed, but a prototype model that conducted smaller calculations was presented to the English government in 1832.

Even though the first machine was never built, Babbage began work on an improved version in 1834. This he called an analytical engine since it allowed the operator to program the type of mathematical operation that the machine performed. The programming was conducted using punch cards, such as those widely used by the textile industry of the 19th century. It is interesting to note that the initial programming of early digital computers in the 20th century used punch cards as well. In many ways Babbage's analytical engine resembled the structure of the modern digital computer. Babbage proposed an input system (punch cards) that supplied data to the unit that actually performed the operations (arithmetic unit). A portion of the machine could be programmed with the operation that the operator wanted to perform, and the results of the calculations were stored in the positions of the gears, which served as a form of temporary memory. As with the digital engine, the output would be supplied to a printing mechanism. While this machine was never built either, a replica was constructed by engineers in 1991 using Babbage's exact engineering specifications. The engine performed the calculations flawlessly. Babbage's analytical engine had foreshadowed the invention of the digital computer by over a century and, in some cases, may have been more precise than the earliest computers.

Babbage's machines were not the only calculating devices proposed during the 19th century. In fact, a number of inventors were experimenting with methods of performing repetitive calculations accurately. One of the more successful of these was the American inventor William Seward Burroughs (1857–1898). Burroughs was the first inventor to successfully attach a keyboard to a calculating machine, which greatly enhanced the usefulness of the machine for the financial and accounting industry. In many ways this device closely resembled the manual adding machines of the 20th century. The operator depressed a series of buttons corresponding to the number to be added. A lever on the side of the machine added (or subtracted) the quantity from the amount in the register. The initial model, presented in 1885, had a number of technical problems, but these were resolved by precision engineering and in 1892 Burroughs obtained a patent for his adding machine. A unique feature of this machine was that it printed the quantities being added and the sum, much the same way as today's electronic adding machine. By the early 20th century Burrough's machine had gained popularity.

The invention of the automated calculating machines of the 19th century was significant primarily for two reasons. First, they represented a technological advance in the processing of repetitive mathematical tasks. These new machines allowed for a greater degree of accuracy and eventually would serve to resolve many of the errors associated with human calculations. Second, these machines would not have been possible without precision engineering skills that were becoming more common by the late 19th century. The link between advances in engineering technology and computer systems continues to this day. In 1946 the first electronic computer, called ENIAC, was invented. Within several decades advances in digital calculating power had resulted in the design of the handheld calculator (1971). Calculators can now be found almost everywhere, and their rapid, accurate ability to perform routine mathematical processes is the cornerstone of our modern society. Engineers are now developing biological computers that use the information storage capabilities of DNA to store vast quantities of data. Still others are investigating the ability to manipulate single atoms to conduct calculations. There can be little doubt that the next generation of computers will be far more powerful and faster than those of today. Yet, they all can trace their calculating ancestry to the inventions of the 19th century.

Selected Bibliography

Bromley, Alan G. *Difference and Analytical Engines*. In *Computing Before Computers*, William Aspray (ed.). Ames, IA: Iowa State University Press, 1990.

Karwatka, Dennis. *Technology's Past, Vol. 2: More Heroes of Invention and Innovention*. Ann Arbor, MI: Prakken Publications, 1999.

Windelspecht, Michael. *Groundbreaking Scientific Experiments, Discoveries and Inventions of the 17th Century*. Westport, CT: Greenwood Publishing, 2002.

Cathode-Ray Tube (ca. 1854–1897): The 19th century was an important time in the development of the atomic theory (see ATOMIC

THEORY, ATOMIC WEIGHT). By the mid-19th century many chemists and physicists had accepted that the atom was the fundamental unit of matter, and in addition many thought that this was the smallest unit of matter that could be detected. At about the same time physicists were making important advances in the study of the electromagnetic spectrum and the properties of electricity (see ELECTRICITY, ELECTROMAGNETIC SPECTRUM). The 19th century invention of the cathode-ray tube (CRT) represents a significant event not only because of the influence that it would have on society in the 20th century, but also because of the information that its invention provided on the nature of the atom and electromagnetism.

In 1645 the German scientist Otto von Guericke (1602–1686) was successful in removing the majority of the air from a sealed tube, creating the first recognized vacuum tube. While this device did not create a true vacuum, it did assist Robert Boyle in his development of the gas laws of the 17th century. Guericke is also recognized for inventing the first static electricity generator, which until the development of the electric cell in the 19th century was the prime method of studying electricity (see ELECTRIC CELL). In the 18th century the American inventor and statesman Benjamin Franklin (1706–1797) concluded from his experiments that electricity flowed from the positive terminal (the **anode**) to the negative terminal (the **cathode**). However, it is now known that electricity flows from cathode to anode (see ELECTRICITY). It was not until the 19th century and the more detailed investigations of the properties of electricity (see ELECTRICITY) that the German inventor Johann Geissler (1815–1869) built a device that could release an electrical discharge inside a vacuum tube. This instrument was commonly called a Geissler tube (ca. 1854). When electricity was applied to the tube, a green glow appeared directly opposite the cathode (negative terminal). The English scientist Michael Faraday had mentioned a similar glow in some of his experiments earlier in the century. At around this time there was a tremendous interest in the study of electromagnetic radiation (see ELECTROMAGNETIC THEORY), and many scientists thought that this glow was the result of electromagnetic radiation within the tube. The German physicist Eugen Goldstein (1850–1930) called these rays cathode rays, because they appeared to have been emitted from the cathode. He named the device that generated them the cathode-ray tube, a name that persists to this day. Goldstein was also one of the first to conclude that Franklin had been wrong in his assessment of the flow of electricity. Instead of flowing from positive to negative, electricity flows from negative to positive.

The English physicist William Crookes (1832–1919) disagreed with the idea that the glow was electromagnetic in nature. Instead, Crookes proposed that it was actually the result of a particle striking the opposite wall of the tube.

To test his hypothesis Crookes made a series of improvements to the Geissler tube, including the use of magnets. Earlier (1858) the German scientist Julius Plücker (1801–1868) had demonstrated that magnets had the ability to deflect the glow within the tube. In his experiments, Crookes first placed a cross-shaped object inside the tube in the path of the suspected particles coming from the cathode. He reasoned that if these were particles moving in a straight line, then their path would be interrupted by the cross. The results of the experiment supported his hypothesis. He also confirmed Plücker's earlier work that magnets could be used to influence the path of the particles. Since the particles were being influenced by an electromagnetic field, it could be concluded that the particles were electrically charged, which effectively ruled out the idea that they were light (see LIGHT).

Crookes and Plücker had demonstrated that the cathode rays could be deflected by a magnetic field, but this did not prove what the particles were. Some suggested that they might be atoms, but the amount of deflection meant that either the particles were exceptionally small or they possessed a very large electrical charge. In 1897 the English physicist Joseph John Thomson (1856–1940) designed an experiment to test which of these ideas was true. From earlier tests Thomson had concluded that the cathode rays could also be bent using electrical charges. By using a specially designed instrument, Thomson was able to measure the amount of deflection with a given amount of charge. From this he was able to determine that their mass was about 1/1000 that of a hydrogen atom (the equivalent of a single proton). What Thomson had discovered were *electrons*, although the name was first applied by George Stoney (1826–1911), an Irish physicist. Electrons are now known to be the fundamental unit of electricity and the basis of all chemical reactions. Thomson's discovery had a tremendous influence on the study of both physics and chemistry in the 20th century, and for his work Thomson received the 1906 Nobel Prize in physics.

Electrons are now known to have a mass that is 1/1840 that of a proton. Their discovery was an important one for the physical sciences. Since the early years of the century scientists had considered the atom to be the smallest unit of matter, and yet the electron had both mass and physical properties. Electrons are utilized not only for moving electrical charges, but also in electron microscopes, which enable scientists to look at very small objects that ordinarily would not be visible using light microscopes. The cathode-ray tube also plays an important role in modern science and society. In the 20th century it was discovered that by using the process first demonstrated by Thomson it was possible to control the movement of the electrons from the cathode. This is the principle that is used in instruments such as **oscilloscopes** and in television. Until recently, computer monitors also primarily used cathode-ray tubes as a display, although these are being quickly replaced by digital displays.

Selected Bibliography

Asimov, Isaac. *Asimov's New Guide to Science*. New York: Basic Books, 1984.

Hornsby, Jeremy. *The Story of Inventions*. New York: Crescent Books, 1977.

Krebs, Robert E. *Scientific Laws, Principles and Theories: A Reference Guide*. Westport, CT: Greenwood Press, 2001.

Parr, G., and O. H. Davie. *The Cathode Ray Tube and Its Applications*. New York: Reinhold Publishing, 1959.

Cell Biology (1817–1898): The study of cells originated in the 17th century with the invention of the first microscopes. Credit for the discovery of the first cells is given to the English scientist Robert Hooke (1635–1701), who described the microscopic compartments of cork cells in 1665. Throughout the 17th and 18th century naturalists described a wide variety of cell types, including human cells such as spermatozoa and red blood cells as well as single-celled organisms such as the **protistans**. In the 19th century a number of important advances were made in the study of cells. By the 1830s the German scientists Matthias Schleiden (1804–1881) and Theodor Schwann (1810–1882) had developed the basic premises of the cell theory and, in the process, established that cells are the smallest unit of life (see CELL THEORY). Shortly thereafter the German biologists Walther Flemming (1843–1915) and Eduard Strasburger (1844–1912) described the process of mitosis, or cell division (see CELL DIVISION). Late in the 19th century Edouard van Beneden (1845–1910), a Belgian biologist, discovered that sex cells undergo a modified form of division, called meiosis, for the production of gametes (see CELL DIVISION). However, 19th century studies of cells were not limited to the description of cellular division. The fact that some cells may be associated with specific diseases resulted in the study of cell **pathology** (see GERM THEORY). In addition, through the use of improved staining techniques (see SYNTHETIC DYES) and enhanced microscopes, biologists were both able to distinguish some of the internal components of cells and examine their metabolic processes. This was effectively the beginning of the science of cell biology.

Several major improvements to the microscope facilitated the 19th century advances in cell biology. Since the invention of the telescope and microscope in the 17th century scientists had been plagued by distortions of the image due to the properties of light as it interacts with the lens. These distortions frequently manifested themselves as color, or chromatic, aberrations. In 18th century an achromatic lens was invented by the English scientist John Dolland (1706–1761) that effectively eliminated this problem. In the 1840s other improvements helped to enhance the magnification capabilities of the microscope. Most notable among these were the invention of the oil-immersion lens by the Italian Giovanni Amici (1786–1863) and the water-immersion lens. These types of lens utilize a layer of liquid between the **objective lens** of the microscope and the sample for the purpose of increasing magnification by focusing more light onto the object. However, despite these advances, cells remain difficult to examine because they are frequently transparent. This problem was solved by the discovery of dyes that stained specific areas of the cell, thus allowing them to be investigated in greater detail.

A number of subcellular compartments, called organelles, were first described in the 19th century. One of the most important of these was the nucleus. Since the nucleus of the cell is the largest organelle and is usually darker in contrast to the rest of the cell, it is highly likely that scientists before the 19th century had observed its presence. However, before the 19th century few people recognized the importance of the cell as the fundamental unit of life (see CELL THEORY) and therefore little attention was directed toward this organelle. This changed in 1831 when Robert Brown (1773–1858), an English botanist, made a series of detailed descriptions of the nucleus. His observations led many to believe that the nucleus was not a cellular anomaly, but rather a fixed component of both plant and animal cells. His work had a tremendous influence on Schleiden, one of the founders of the cell theory (see CELL THEORY). Schleiden mistakenly believed that new cells were formed by a budding of material from the nucleus. The discovery of the process of mitosis several years later corrected this viewpoint (see CELL DIVISION). Perhaps the most important contribution of the nucleus to modern science was not in the organelle itself, but rather its contents. The nucleus contains the genetic material of the cell, called deoxyribonucleic acid (DNA). DNA was first discovered in the cell in 1869 by Johann Miescher (see BIOMOLECULES), although 19th century biologists did not recognize its role as the hereditary material. In 1882 Flemming used the term *chromatin* to identify the mass of material inside the nucleus. A few years after this the German scientist Heinrich Wilhelm Waldeyer (1836–1921) identified individual structures within the chromatin called chromosomes (see CELL DIVISION). Today, biologists classify cells into two broad categories depending on the presence or absence of a nucleus. Primitive cells that lack a nucleus are called **prokaryotic** cells, and cells that contain a nucleus are called **eukaryotic** cells. Plant and animal cells are both examples of eukaryotic cells.

Until the invention of staining the majority of studies of cells involved the use of plant cells since the internal components of these cells were easier to distinguish. One of the main traits distinguishing plant and animal cells is the presence of the predominantly green photosynthetic pigments in plant cells. By the 19th century scientists had determined that plants require carbon dioxide, water and sunlight to manufacture sugars and that oxygen was a by-product of this process. In the 19th century the discoveries primarily focused on the location of these chemical reactions. The green pigment of plants responsible for the process of photosynthesis was first isolated in 1817 by two French industrial chemists, Joseph Caventou (1795–1878) and Pierre Pelletier (1788–1842). They named the compound chlorophyll after the Greek words for "leaf" and "green." In 1837 the French scientist Henri Joachim Dutrochet (1776–1847) discovered that chlorophyll was necessary for plants to manufacture sugar. For several decades it was believed that the chlorophyll was evenly distributed around the interior of the cell. However, around 1862 Julius von Sachs (1832–1897) proposed that there were actually small organelles that contained the photosynthetic pigments. The term *chloroplast* was later adopted as a name for these structures. In 1893 the term *photosynthesis* was first used to explain these linked chemical reactions. Since that time biologists have worked out the major metabolic pathways of photosynthesis. In most photosynthetic organisms an initial phase coverts the energy of the sun into chemical energy, followed by a second set of reactions that assembles sugars, starches and other biomolecules. The 20th century understanding of these processes was made possible by the 19th century botanists who identified the chloroplast as the site of the reactions.

Several other cellular components were discovered during the 19th century (see Figure 5). As part of his definition of the cell theory, Schleiden developed the terms *nucleoplasm* and *protoplasm* to identify the fluid areas inside and outside the nucleus, respectively (see CELL THEORY). The protoplasm is now commonly called the **cytoplasm**. Two additional organelles were discovered before the end of the century. The first of these discoveries was the result of a new staining technique developed by the Italian biologist Camillo Golgi (1844–1926). Golgi had a strong interest in the study of neuroscience and developed a new staining method that used silver salts to examine the structure of nerve cells. During his studies (ca. 1892–1898) he described a cellular organelle that appeared to be very important to cell function. This organelle was later called the Golgi body (sometimes also called the Golgi apparatus or Golgi complex) after its discoverer. Golgi was the recipient of the 1906 Nobel Prize in physiology or medicine for his work in the neurosciences. The Golgi bodies are now recognized as the processing centers of the cell, and as such are responsible for the processing of organic molecules for both export and internal use. In 1898 another important organelle was discovered by a German biologist named Carl Benda (1857–1933). This organelle was the mitochon-

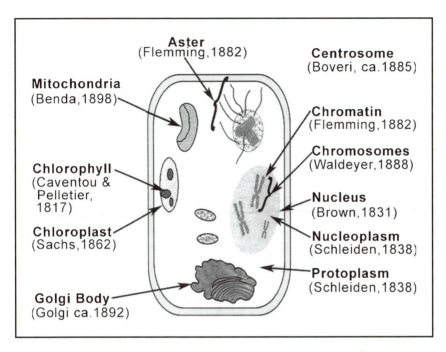

Figure 5. The major structures of the plant cell discovered during the 19th century, along with the principal discoverer and year of discovery. With the exception of chlorophyll and the chloroplast, similar structures were found and studied in animal cells.

dria, and although 19th century biologists did not understand its function, it is now classified as the powerhouse of the cell. This is the site where carbohydrates, fats and proteins are converted to energy by the process of **cellular respiration**. This process was partially worked out by Hans Krebs (1900–1981) around 1937. By the end of the 19th century a fairly detailed picture of the interior of the cell was developing.

It should be noted that the study of cell biology during this time was not limited solely to the identification of internal organelles. At the same time as these discoveries an interest was developing in understanding the metabolic processes of the cell. For centuries scientists recognized that biological organisms convert raw materials into new chemical structures. One of these processes, fermentation, attracted a strong interest of the scientific community due to its economic importance in producing alcoholic beverages. In the 19th century biologists such as Louis Pasteur attempted to explain the process of fermentation as a cellular process (see FERMENTATION, GERM

THEORY). Others took a more chemical approach and in the process established the groundwork for a later understanding of metabolic catalysts called enzymes (see ENZYMES).

While the study of cell biology effectively began during the 19th century, the major discoveries of this field would occur in the second half of the 20th century. Today's cell biologists not only explore the biochemical reactions occurring within the cell, but some also are actively looking for evidence of cells at other locations within the solar system. Today's cell biologists utilize new molecular and biochemical techniques to manipulate the chemical processes of the cell in an attempt to understand disease. The recent advances in this field have made it possible to convert some cells into microscopic biological factories capable of producing molecules of interest. All of these achievements are the result of the strong foundation of knowledge in cell biology established in the 19th century.

Selected Bibliography

Magner, Lois N. *A History of the Life Sciences*, 2nd ed. New York: Marcel Decker, 1994.
Mayr, Ernst. *The Growth of Biological Thought: Diversity, Evolution and Inheritance.* Cambridge, MA: Harvard University Press, 1982.
Serafini, Anthony. *The Epic History of Biology.* New York: Plenum Press, 1993.
Singer, Charles. *A History of Biology to about the Year 1900: A General Introduction to the Study of Living Things.* Ames, IA: Iowa State University Press, 1989.

Cell Division (1830–1888): With the exception of the uncertain origin of the first cells, modern biologists recognize that all cells arise from preexisting cells. However, this belief was not always the case. Using one of the early microscopes, which were little more than magnifying lenses, the English scientist Robert Hooke discovered the first cells in the 17th century. Although a number of different cell types were discovered over the course of the next century, most scientists at the beginning of the 19th century considered that tissues were the fundamental unit of living organisms, and thus very little attention was directed at understanding the process of cell division. Various investigators had provided descriptions of dividing cells during the 17th and 18th centuries, but the meaning of the process was unknown. However, in the early decades of the 19th century a number of discoveries initiated an interest in more detailed studies of cells. This included the development of more powerful microscopes, which in turn led to the discovery that the majority of cells contain a nucleus (see CELL BIOLOGY). The development of the cell theory by the German biologists Matthias Schleiden and Theodor Schwann in the 1830s strongly suggested that the cell was the prime unit of life (see CELL THEORY). The formation of the cell theory marked the beginning of a dedicated interest among biologists to understand the process of cell division. In doing so, the investigators of cell division had to work through a number of differing ideas and misconceptions surrounding the genesis of life.

During the development of the cell theory, Schwann proposed that all life originated as individual cells. He reasoned that because many living organisms are multicellular, some mechanism must exist by which these cells reproduce. From the English scientist Robert Brown's observation of the nucleus (1831), Schleiden in 1838 suggested that new cells formed from a budding of material from the nucleus, but he also contended that it was possible for cells to develop from an inorganic fluid called the cytoblastema. The modern terms *protoplasm* and *cytoplasm* are now used to describe this fluid (see CELL BIOLOGY). So great was the power of the cell theory that for a while the concept was accepted by the majority of the scientific community. However, over the next few decades several investigators began to doubt that cells could form in this manner. One of the leaders of this next generation of cell biologists was Karl Nägeli (1817–1891), who studied cell division in algae. Nägeli's observations did not support the idea of spontaneous generation and led him to believe that all cells arose only from preexisting cells. Similar studies in the 1850s of the embryonic division of eggs from organisms such as the frog, cuttlefish and rabbit by Robert Remak (1815–1865), Albert Kolliker (1817–1905), and Martin Berry (1802–1855) provided additional evidence that cells did not arise spontaneously. However, the final blow to the Schleiden's idea that cells may be formed directly from the protoplasm came from the work on cells by the German scientist Rudolf Virchow (1821–1902). An influential biologist and leading investigator of cell pathology, Virchow published several important papers that refuted the idea of spontaneous generation.

By the mid-19th century most biologists had accepted the cellular basis of life, although the mechanism of cell division still remained elusive. During the 1870s a number of improvements were made in microscopic techniques, most notably the use of cytological stains that had the ability to distinguish the internal components of the cell in greater detail (see SYNTHETIC DYES). One of the pioneers in the use of cellular staining techniques was the German scientist Walther Flemming. Kolliker had previously noted that the nucleus of a cell disappeared completely while the cell was dividing, only to reappear in both of the new cells following division. This phenomenon served to focus attention on the nucleus during cell division. Using his new staining techniques (ca. 1882), Flemming was able to distinguish and classify the several phases of cell division in animals, a process he called *mitosis* due to the thread-like nature of the structures formed during division. These types of studies that used stains to examine cellular structures were greatly enhanced by the invention of synthetic dyes and stains in the 19th century (see SYNTHETIC DYES). Another German biologist, Eduard Strasburger, was doing similar work in plants at around the same time. Together these scientists developed much of the terminology still used to describe the process of mitosis.

The stains used by Flemming and Strasburger had the ability to detect small structures within the nucleus. Flemming labeled this mass of material the *chromatin*. Several years later the German biologist Heinrich W. Waldeyer used the term *chromosomes* to indicate the individual pieces of the chromatin, although a number of earlier investigators had indicated their presence during division. The process of cell division (see Figure 6) involves a reorganization of the chromatin and chromosomes (see Table 1). This is done by a thread-like structure that Flemming called an *aster*. A few years later (1888), Theodor Boveri (1862–1915) discovered that the threads of aster (now called spindle fibers) were controlled by a structure he called a centrosome (now called a centriole). Strasburger's studies provided the current names for the individual stages of mitosis. Table 1 lists the events of what 19th century researchers recognized was occurring during each of these stages. Although biologists provide labels for these phases, the process of mitosis is a continuous process in most cells. While the process of nuclear division is the same for plants and animals, the division of the cytoplasm at the end of mitosis (called cyto-kinesis) differs slightly between plants and animals. Plant cells use specialized **vacuoles** within the cell to reform the cell wall, thus splitting the cell in two. Animal cells use part of the **cytoskeleton** to constrict the cell membrane inward, forming two new daughter cells.

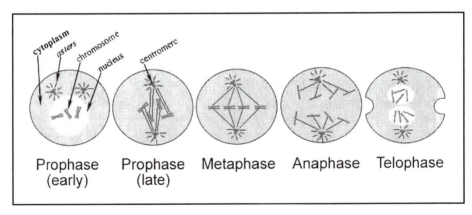

Figure 6. The stages of mitosis. This diagram depicts an animal cell undergoing cell division. Plant cells differ only in the last stage (see text).

Table 1
The process of mitosis as known to 19th century biologists. The modern terms for structures are provided in parentheses. Modern cell biologists frequently break prophase into an early and a late (prometaphase) stage

Stage	Events
Prophase	Chromosomes may be distinguished by stains
	Nuclear membrane disappears
	Centrosome (centriole) and aster (spindle fibers) form
	Centrosome and asters move to opposite sides of the cell
	Fibers from aster connect to chromosomes
Metaphase	Chromosomes align along a central line in the cell
Anaphase	Chromosomes split and head to opposite sides of the cell
Telophase	Nuclear membrane reforms around the chromosomes
	Protoplasm (cytoplasm) splits in animals, forming two new cells
	In plants, a new cell wall forms, splitting the two new cells

The discovery of the process of mitosis was an important one for the biological sciences for several reasons. Not only did it provide a final answer to the question of how cells divide, and thus provide validity to the cell theory (see CELL THEORY), but it also stimulated an intense interest in understanding the biological process of inheritance (see INHERITANCE). However, mitosis is not the only mechanism by which cells divide. Prokaryotic cells, those lacking a nucleus, frequently divide by **binary fission**, and some yeast cells divide by a form of budding. In 1887 Edouard van Beneden, a Belgian biologist, recognized that the process of mitosis did not occur in exactly the same manner during sexual reproduction. Beneden was the first to understand that all cells of the body contain the same number of chromosomes but that, during the process of forming egg and sperm cells (gametes), the number of chromosomes is reduced by half. This process is called meiosis, and the study of its relationship to the chromosomal basis of inheritance (see INHERITANCE) occupied a large part of the early 20th century research in genetics.

The study of mitosis and meiosis was not completed during the 19th century. For example, mitosis is now recognized to be just one portion of the cell cycle, which includes all of the active phases of a cell's life. The study of cell cycle regulation is a major focus of current research into the causes of **cancer**, which in many cases involves the unrestricted mitotic division of mutated cells.

The process of meiosis, discovered at the end of the 19th century, is now known not only to include a segment that reduces the number of chromosomes in the sex cells, but also to contain a phase that introduces variability into the **genome** by shuffling the genes and chromosomes of the parents and randomly assigning them to the gametes. The chromosomal basis of this variation helps to explain the observations of 19th century evolutionary biologists (see HUMAN EVOLUTION) as well as providing the foundation for fields of biology such as **molecular biology** and **population genetics**.

Selected Bibliography

Magner, Lois N. *A History of the Life Sciences*, 2nd ed. New York: Marcel Decker, 1994.

Mayr, Ernst. *The Growth of Biological Thought: Diversity, Evolution and Inheritance*. Cambridge, MA: Harvard University Press, 1982.

Serafini, Anthony. *The Epic History of Biology*. New York: Plenum Press, 1993.

Singer, Charles. *A History of Biology to about the Year 1900: A General Introduction to the Study of Living Things*. Ames, IA: Iowa State University Press, 1989.

Cell Theory (1831–1866): The term *cell* is used to represent the smallest living unit that has the characteristics of all of life, meaning that it has the ability to respond to its environment, reproduce, and conduct metabolic processes. The term itself first originated in the 17th century when the English scientist Robert Hooke used it to describe the honeycomb arrangement of compartments found within dead cork tissue. Using the primitive microscopes of the 17th century, microbiologists such as the Dutch scientist Anton van Leeuwenhoek (1632–1723) were able to identify individual spermatozoa and protozoan cells. Despite an interest in the microscopic world, the microbiologists of the 17th and 18th centuries did not recognize that all living things were composed of cells. The actual study of cell biology began in the 19th century, when improvements in microscope design allowed for a more detailed examination of cell structure. The study of cells in the 19th century included not only the formal definition of the cell theory, or the basic properties of cells, but also the study of cell division (see CELL DIVISION) and some introductory descriptions of the internal organelles of cells (see CELL BIOLOGY).

At the start of the 19th century the fundamental unit of life was considered to be tissues. This was primarily because of the work of the French anatomist Xavier Bichat (1771–1802), who defined twenty-one different tissue types on the basis of their physical characteristics. Bichat recognized that organs were composed of combinations of these tissue types and that organs were then organized into organ systems that had distinct functions in the body, for example, the respiratory or skeletal system. Furthermore, Bichat explained how tissues responded to stimuli, thus further supporting his ideas that they were the fundamental unit of life. Although Bichat was incorrect in this regard, the strength of his research influenced biologists for several decades. Bichat's work is considered to be the starting point for the study of **histology**, or the study of tissue structure.

By the 1830s, additional information was being accumulated that began to cast doubt on tissues being the organizational unit of life. The majority of this information was being derived from the study of plant cells. The study of plant cell structure had begun in the 17th century with the work of Marcello Malpighi (1628–1694) and Nehemiah Grew (1641–1712). Both of these scientists prepared drawings of the internal structures of plant cells, but did not draw any conclusions as to their function. However, in 1831 the English botanist Robert Brown made a series of detailed observations on the nucleus of plant cells (see CELL BIOLOGY). Other scientists before him had noticed the cell nucleus, but Brown is considered to have been the first to determine that most cells contained a nucleus and is credited with developing the term *nucleus*. Brown's work had a tremendous influence on another botanist, Matthias Schleiden.

Schleiden almost immediately recognized the importance of Brown's discovery. As a botanist Schleiden was not content with the classification of plants, a major area of research in early 19th century botany. Instead, Schleiden wanted to understand the structure and internal organization of plants. In 1838 Schleiden contended that while animals are viewed as individuals, plants are actually groups of individuals compiled into one organism. Schleiden proposed that cells were these individual components of plants and that the nucleus that Brown had described was an important factor in the development of the cell. Although Schleiden was mistaken in thinking that new cells formed through a process of budding from the nucleus of existing cells, he was correct in deducing the cellular basis of life. Later studies would more accurately describe the process of cell division (see CELL DIVISION). Schleiden's work was important not only for this recognition cell's role as the unit, but also for stimulation of research into the microscopic structure of plant cells. Schleiden made many important discoveries about plant structures and influenced an entire generation of scientists dedicated to microscopic studies (see CELL BIOLOGY)

Schleiden's ideas on cells caught the attention of another German biologist, Theodor Schwann. Schwann also made a number of important contributions to 19th century biology (see EMBRYOLOGY), but he is most frequently credited for relating Schleiden's ideas on plant cells to animals. Schwann recognized that, despite the differences between plant and animal cells and the diversity of cell structure in tissues, a common factor was present in all cells. That structure was the nucleus. In his microscopic examinations of the **notochord** and cartilage of tadpoles, Schwann determined that all animal tissues are composed of cells and that all an organism's structures originate from these cells. He further proposed that in multicellular organisms the cells are under the control of the organism. These three points form the basis of the cell theory. Schwann's ideas were presented in 1839 in his publication *Microscopical Researches into the Accordance in the Structure and Growth of Animals and Plants.* Figure 7 is an example of the cells that Schwann described in this work. This work is also important in its conclusion that the process by which cells originate is the same in both plant and animal cells. This idea provided an important biological connection between the plant and animal kingdoms. While Schwann was also correct in his conclusion that cells were living organisms, his thoughts on cell division were similar to those of Schleiden (see CELL DIVISION).

Following the work of Schleiden and Schwann an intense interest developed in the microscopic basis of life. Much of this work centered on the studies of cellular structures (see CELL BIOLOGY), cell division (see CELL DIVISION) and the development of new organisms (see EMBRYOLOGY). However, the cell theory presented by Schwann and Schleiden also helped to explain the nature of organisms such as protozoa, which are single-celled creatures. In 1866 the German biologist Ernst Haeckel (1834–1919) proposed a new classification for single-celled organisms, the protistans. Today, biologists recognize an entire kingdom of these groups, which are classified by their single-celled nature and the presence of a nucleus. Other scientists, such as the German scientist Rudolf Virchow utilized the principles of the cell theory to establish the relationship between cells and disease. Virchow's work is considered by many to be exceptionally influential in the initiation of the study of cell pathology.

There can be little doubt as to the importance of the development of the cell theory during the 19th century. Modern cell biologists study cells from many perspectives, including the great biodiversity of protistans and the molecular basis of cell function. A large number of single-celled organisms are responsible for diseases in humans. Physicians and researchers now recognize that almost all human ailments have a cellular basis. The establishment of the cell theory, along with the formation of the germ theory of disease (see GERM THEORY), played a major role in the development of modern medicine and the field of cell biology. These types of studies, with their implications for advances in medicine, agriculture and an understanding of life on our planet, are a direct result of the 19th century discovery of the cell theory by Schleiden and Schwann.

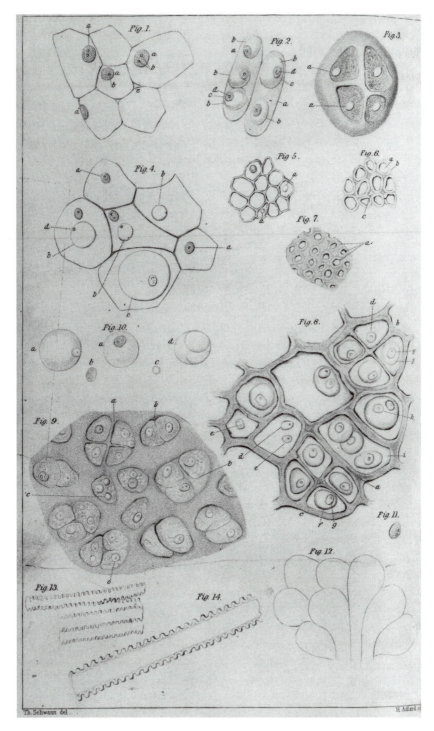

Figure 7. An example of the cells that Schwann studied during the 19th century. From *Microscopical Researches into the Accordance in the Structure and Growth of Animals and Plants*, 1839. (Library of Congress photo collection)

Selected Bibliography

Bynum, W. F., E. J. Browne and Roy Porter (eds.). *Dictionary of the History of Science*. Princeton, NJ: Princeton University Press, 1981.
Magner, Lois N. *A History of the Life Sciences*, 2nd ed. New York: Marcel Decker, 1994.
Mayr, Ernst. *The Growth of Biological Thought: Diversity, Evolution and Inheritance*. Cambridge, MA: Harvard University Press, 1982.
Singer, Charles. *A History of Biology to about the Year 1900: A General Introduction to the Study of Living Things*. Ames, IA: Iowa State University Press, 1989.

Chemistry (ca. 1810–1852): From a historical perspective the science of chemistry is relatively young. In the 17th century the work of the pioneer English chemist Robert Boyle began to separate the study of chemistry from the practice of **alchemy**. However, it was not until the 18th century and the investigations of the French chemist Antoine Lavoisier that modern chemistry can be said to have begun. Lavoisier developed the practice of quantitative chemistry, which involves the study of chemical reactions based on the collection of data from experiments, and not solely on philosophical theories. In taking this approach he established some of the fundamental principles of chemistry. Many of them served as the basis for several important developments of the 19th century, such as the identification of new elements (see ELEMENTS), the establishment of atomic weights for elements and compounds (see ATOMIC WEIGHTS) and research into the nature of the atom (see ATOMIC THEORY). Yet one of the greatest achievements in the study of chemistry during this time originated as a series of explanations of how elements combine to form compounds.

Before the formation of an organized table of the elements, many chemists had noted that in chemical reactions some groups of elements reacted similarly. This discovery played an important role in the development of the Periodic Table of the Elements in the early 19th century (see ELEMENTS). Many early studies of chemical reactions involved the study of oxides, or compounds containing oxygen, and the term *radical* was used to denote the substance interacting with the oxygen molecule. Several explanations were proposed as to why this may occur. The renowned Swedish chemist Jöns Berzelius suggested that the combination of elements was to the result of what was called electrochemical dualism. Berzelius, as well as others, suggested that all elements have a distinct charge and that compounds are formed when negatively and positively charged substances combine, thus neutralizing the charge. While it is now recognized that these electrical differences are responsible for a class of chemical bonds called **ionic** bonds, which are found frequently in salts, they do not explain the majority of chemical bonds, especially those involving carbon. In modern chemistry this latter group is called covalent bonds and represent a sharing of valence electrons between atoms. However, the discovery of electrons would have to wait until later in the 19th century.

By the mid-1800s additional information was accumulating to suggest that chemical reactions were not necessarily based solely on electrochemical differences. In 1853 the French chemist Jean Baptiste Dumas noticed that when chlorine reacted with **hydrocarbons**, the chorine replaced the hydrogen in the molecule, despite chlorine being negative and hydrogen being positive. This was called the process of substitution. Although this discovery was made by Dumas, the main proponent of the theory was the French chemist Auguste Laurent (1808–1853). This process of substitution played an important role in the early 19th century studies of organic chemistry (see ORGANIC CHEMISTRY). Substitution still relied on the idea of radicals, although these radicals now represented the core portion of the molecule in the reaction and not individual atoms of an element. Dumas called these core components *types*, and thus the substitution theory is sometimes called the theory of types.

The next major development in this area was the result of a series of investigations performed by the French-German chemist Charles Gerhardt. Gerhardt studied reactions involving organic compounds, or compounds that contain carbon. He noted that when organic molecules reacted with one another, inorganic material such as water frequently remained. Gerhardt called this leftover material *residues*, and he proposed that the formation of residues could explain the nature of chemical reactions. Gerhardt then revisited the theory of types and suggested that organic materials could be classified according to the types of residues that they formed. He arranged compounds into what are called *homologous series*, based on their chemical reactivity. What Gerhardt had recognized was that chemical reactivity of an organic compound is determined by the **functional groups** attached to the carbon chain. There are a large number of functional groups, but each gives the compound a unique characteristic. For example, all alcohols contain a hydroxyl functional group (–OH). Understanding the actions of these functional groups is the basis of the study of organic chemistry (see ORGANIC CHEMISTRY).

Compounds that are part of a homologous series display the same number of substitutions in a chemical reaction. This is true not only for compounds, but for elements as well. Using the single substitution of hydrogen and chlorine as a reference, compounds were identified by the number of substitutions that occurred during a chemical reaction. In 1852 the English chemist Edward Frankland (1825–1899) published the results of his studies on a group of chemical compounds called organometallics. Frankland had noticed that these compounds, which have metallic ions integrated into organic structures, always combined in certain ratios. This fixed value was originally called the combining power and is similar in nature to the substitution concept mentioned above. The term was changed to *valence* in the late 19th century. The idea of valency had a tremendous impact on the study of chemistry later in the century. At the atomic level, the valence value of an element is now known to be dependent on the number of electrons in the outer energy shell of an atom. Frequently called the valence electrons, they are responsible for not only the chemical reactivity of an atom, but also the relationships among the elements that enabled the construction of the early periodic tables of the elements (see ELEMENTS for figure).

The electron as an atomic particle would not be discovered until 1897 by the English physicist Joseph John Thomson (1856–1940), although its existence had been theorized several years earlier. The understanding that valency is a function of electron arrangement was not established until the early 20th century by a number of chemists involved in the study of **electrochemistry** (see also ELECTRIC CELL). However, in the 19th century chemists now had a mechanism for predicting and characterizing chemical reactions, even if initially they did not understand the basis of these relationships. The identification of the valence structure of organic compounds began with the work of Friedrich Kekulé (1829–1896). Kekulé was the first to suggest that carbon was tetravalent, or had the ability to form chemical bonds with four other chemical substances. Once this was established, the science of organic chemistry and the identification of organic structures commenced (see ORGANIC CHEMISTRY). However, Kekulé's work was the result of a dedicated attempt during the first half of the 19th century to understand the nature of chemistry. The 19th century theories on the valence structure of atoms are now commonly taught in introductory science classes, where they are a fundamental principle on the organization of matter. The valence structure of an element or compound was the culmination of the theories of Berzelius, Gerhardt, Dumas and a large number other experimental chemists of the 19th century.

Selected Bibliography

Cobb, Cathy, and Harold Goldwhite. *Creations of Fire: Chemistry's Lively History from Alchemy to the Atomic Age*. New York: Plenum Press, 1995.

Nye, Mary Jo. *Before Big Science: The Pursuit of Modern Chemistry and Physics 1800–1940*. New York: Twayne Publishers, 1996.

Russell, C. A. *The History of Valency*. New York: Leicester University Press, 1971.

Taton, Rene (ed.). *History of Science: Science in the Nineteenth Century*. New York: Basic Books, 1965.

E

Earth (1835–1886): The 19th century was a time of tremendous advances in the field of astronomical science. Over the course of this century astronomers developed the means to determine the distance to stars and their relative motion to one another (see ASTRONOMY). Studies of our Sun indicated that it was not the perfect sphere suggested since the time of Aristotle, but rather a dynamic structure that influenced the magnetic field of the Earth (see THE SUN). The discovery of new planets, moons and asteroids further indicated that our solar system was a complex place, filled with more objects than had previously been thought (see PLANETARY ASTRONOMY). However, until this time the Earth had primarily been used as an observational platform to observe the solar system, and little had been done to explain the planetary nature of our home planet. This would change with a series of significant discoveries in the 19th century.

One of the troubling issues for scientists since early times was a method for estimating the age of the Earth and solar system. For most of history these estimates had relied on religious records, which in the Western world focused on studies of the Bible. The Catholic (English) bishop James Ussher (1581–1656) had used the biblical record to establish that the Earth was formed in 4004 B.C.E., and thus was less than 6,000 years old. Although this idea lacked any scientific validity, it remained one of the primary estimates until the 19th century. In 1854 Hermann von Helmholtz (1821–1894) and William Thomson (also known as Baron Kelvin) suggested that it was possible to estimate the age of the Sun by examining its constriction in size due to gravity (see SUN). Using this method they were able to determine that the solar system, including the Earth, was approximately 25 million years old. While this estimate may seem to be an improvement over earlier biblical calculations, it presented significant problems for geologists and biologists, who

required a greater expanse of time for their developing theories (see GEO-
LOGICAL TIME, GEOLOGY, HUMAN EVOLUTION, THEORY OF
ACQUIRED CHARACTERISTICS, THEORY OF NATURAL SELEC-
TION).

The fact that the Earth rotates on its axis is common knowledge in our
modern world. However, for most of recorded time it was believed that the
stars, moons and planets all revolved around the Sun. The 16th century Polish
astronomer Nicholas Copernicus is recognized as widely developing the idea
that our solar system is heliocentric, or sun-centered. In the 17th century
Galileo championed the concept with his evidence that the newly discovered
moons orbiting around Jupiter resembled a small solar system. Because our
solar system behaved in much the same way, he reasoned that the movement
of the objects in the sky must be due to the rotation of the Earth. However,
until the 19th century this rotation had never been demonstrated scientifically
and instead had relied on observation only. This belief changed in 1851 when
Jean Leon Foucault (1819–1868) experimentally measured the rate of the
Earth's rotation. Foucault recognized that the motion of a pendulum was
always in the same direction or plane, even when the platform to which it was
attached was placed in motion. Foucault's experimental system consisted of a
60-pound iron ball suspended 200 feet from the dome of a church. The end
of the iron ball was fitted with an iron spike. By covering the floor of the
church with sand, Foucault was able to track the movement of the Earth by
the markings made in the sand by the pendulum over time. The rate of this
motion varied with latitude.

Foucault's experiment set the stage for one of the most important 19th
century discoveries regarding the motion of the Earth. In 1881 the American scientist Albert Michelson (1852–1931) invented an instrument called an
interferometer. The interferometer functions by passing a beam of light
through a piece of glass half covered in silver. The light striking the upper
portion of the glass is reflected, but the light striking the lower half passes
through the glass (see Figure 8). Both beams are then reflected through a series
of mirrors until they are recombined at a single point. In principle if the split
beams traveled at exactly the same speed then the light at the combination
point would be unchanged from the original. However, if one of the beams
were slower, it would arrive at the final point out of phase, producing what
is called an interference fringe. A similar pattern had been demonstrated by
the English scientist Thomas Young (1773–1829) earlier in the century in his
studies of light (see ELECTROMAGNETIC SPECTRUM, LIGHT).

Between 1881 and 1886 Michelson and his assistant Edward Morley
(1838–1923) used this instrument in an attempt to detect the presence of
ether. The theoretical compound ether had existed in many forms since the

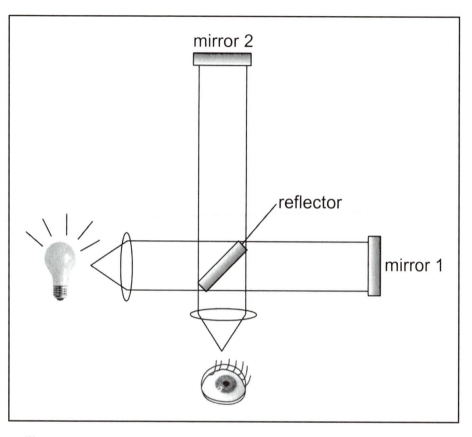

Figure 8. An example of the interferometer designed by Michelson in 1881. This instrument was used to disprove the concept of ether in the 19th century (see text for details).

Scientific Revolution. Before the development of the atomic theory, ether was believed to be the compound that occupied the space between atoms. In the 19th century it was thought to be the substance that was responsible for transmitting light and other electromagnetic radiation from the Sun. Many physicists believed that light was similar in nature to sound and therefore required a medium to be transmitted across a distance (see LIGHT). Michelson and Morley decided to search for the presence of ether by orientating their interferometer so that the original beam of light was parallel to the Earth's motion. This orientation meant that the reflected beam should be slowed by the ether and thus arrive out of phase from the original beam. Despite repeated attempts, the beam always arrived in phase. Although the Michelson–Morley experiments set out to test the presence of ether, they effectively provided the final evidence needed to disprove its existence.

Studies of the Earth in the 19th century were not limited to the astronomical sciences. The early 19th century is considered to be the Heroic Age of Geology, during which many important discoveries were made and theories developed (see GEOLOGY, GEOLOGICAL TIME, PALEONTOLOGY). Furthermore, studies of weather are directly connected to an understanding of the Earth, and in the 19th century this led to a number of important discoveries in the field of meteorology, including the Coriolis effect, which explains the motion of air masses in each hemisphere (see METEOROLOGY). The Earth remains a model system in astronomical studies. Because it is the only planet on which we currently have evidence of life, studies of the Earth's distance from our Sun have provided guidelines in the search for life-bearing planets in other solar systems.

Selected Bibliography

Asimov, Isaac. *Asimov's New Guide to Science*. New York: Basic Books, 1984.

Gohau, Gabriel. *A History of Geology*. London, England: Rutgers University Press, 1990.

Krebs, Robert E. *Scientific Laws, Principles and Theories: A Reference Guide*. Westport, CT: Greenwood Press, 2001.

Spangenburg, Ray, and Diane K. Moser. *The History of Science in the Nineteenth Century*. New York: Facts on File, 1994.

Electric Cell (1800–1833): Electric cells may be considered the 19th century precursor to the modern electric battery, which is simply a combination of several individual cells. The invention of electric cells in the early 19th century followed two centuries of preliminary scientific studies of electricity (see ELECTRICITY). As a natural force, electricity in the form of lightning from thunderstorms is one of the more visible forces of nature, and thus there was a natural scientific interest in understanding its action. The study of electricity before the 19th century generally focused on static electricity. However, by the start of that century a number of scientists had begun to study the motion of electricity, frequently called dynamic electricity.

Since the 17th century, and the inventions of the German scientist Otto von Guericke, it had been possible to generate static electricity. The invention of the first practical device to hold an electrical charge was the Leyden jar. The Leyden jar was invented in 1745 by the German scientist Ewald Georg von Kleist (1700–1748), although similar inventions were in existence around the same time. The Leyden jar (see Figure 9) functioned as an early form of **capacitor** in that it temporarily stored an electrical charge, which was then discharged in a single instant. The design was later improved upon by other scientists. While it was a useful invention for the study of static electricity, the

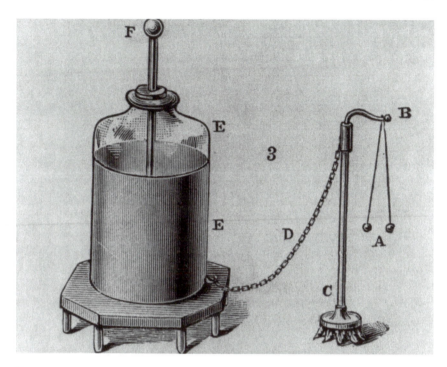

Figure 9. An example of a Leyden jar from the late 18th century. This device was used to temporarily store electrical charges for the study of electricity. (Hulton Archive/Getty Images)

Leyden jar did not provide the sustainable, consistent electrical current needed for dynamic studies.

The invention of the first electric cell to deliver a constant current began with the study of electrical currents in animals during the 18th century. The Italian physiologist Luigi Galvini (1737–1798) conducted a number of detailed experiments on animal electricity starting around 1750. While performing a frog dissection Galvini noticed that the muscles of the frog contracted when touched with a brass instrument, but only when the frog was lying on an iron plate. Galvini conducted a large number of experiments to define animal electricity, but he mistakenly believed that it was generated directly by the muscle. However, his experiments were important because they attracted the attention of the Italian physicist Alessandro Volta (1745–1827).

Unlike Galvini, Volta believed that the observed phenomenon of animal electricity was caused by the metal of the instruments. Around 1800 he set out to experimentally prove his ideas. First he replicated Galvini's experiments with a variety of metals, except that instead of using animal muscles to indicate the presence of electrical currents he used an instrument of his own design called an electroscope. The electroscope was a much more sensitive instrument than the muscle tissue and allowed Volta not only to quantify the electrical charge produced by different combinations of metals, but also to record whether the metal was left with a positive or negative charge following an experiment. Volta also recognized that liquids, especially acids, played a role in electrical conduction. Without liquids, metals quickly dispersed their electrical charge, but when the two dissimilar metals were separated by a weak acid the resulting charge had a significantly longer duration. Using this information Volta constructed the first wet electric cell in early 1800. The device, sometimes called a voltaic pile or electric pile, was made of a series of stacked zinc and copper (or silver) plates separated by a liquid (see Figure 10). Volta experimented with the design of a number of these piles, substituting various metals and liquids to get the greatest electrical charge.

Unlike the Leyden jar, which simply stored an electrical charge, Volta's electric cell actually generated an electrical current. Furthermore, the voltaic

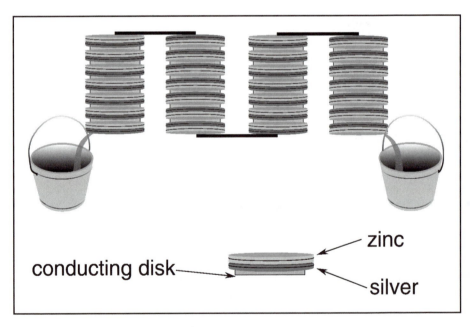

Figure 10. An illustration of a 19th century voltaic pile first designed by Alessandro Volta in 1800. The conducting disks between the metal plates were soaked in a weak acid to increase conductivity.

pile had the ability to deliver a constant electrical current, unlike the rapid static discharge of the Leyden jar. The many applications of the voltaic pile were quickly recognized by the scientific community, and a number of improvements were made to the device over the next several decades. The original battery required a large quantity of liquid, but as the liquid evaporated the battery lost its effectiveness. To remedy this problem Volta constructed a battery that used cardboard separators between the metal plates. These separators were soaked in an acidic solution thus allowing the conduction of electricity. Other advancements followed, each improving not only the amount of current, but also reliability of the electric cell to provide a sustainable current.

The electric cell was quickly adopted by the chemists of the early 19th century as an improved method of delivering an electrical current to study the elemental components of chemical compounds. In 1800 the English scientist William Nicholson (1753–1815) noticed that while his electric cell was in operation tiny gas bubbles emerged from the submersed wires. These gases were hydrogen and oxygen, the elements of water. Nicholson had demonstrated that it was possible to break chemical bonds using electricity, a process called *electrolysis*. This was the beginning of the field of electrochemistry. In 1807 another English scientist named Humphrey Davy designed a powerful electric cell consisting of more than 500 plates. This cell produced enough current to separate the potassium from potassium carbonate by electrolysis. This method proved to be very successful, as Davy was able to isolate four more elements within the next few years (see ELEMENTS). Similar electrical studies were also being performed on chemical compounds, resulting in a field of study called electrochemistry. While scientists were using electrolysis early in the century, they did not fully understand how electrolysis worked. In 1833 Michael Faraday, an English pioneer in the understanding of both electricity and electromagnetic theory (see ELECTRICITY, ELECTROMAGNETIC THEORY), explained the process. In doing so he developed the terms *anode, cathode* and **electrolyte** to explain the operation of the battery.

The electric cell was an exceptionally important advance for early 19th century physicists. It provided a reliable power source for experiments, but it also was quickly adapted for industrial use. The electric cell would play an important role in the later invention of the telegraph (see TELEGRAPH) and the development of the principles of electricity (see ELECTRICITY). It also found use as an ignition source in the internal combustion engine (see INTERNAL COMBUSTION ENGINE). Modern society uses electric batteries in every aspect of daily life. Although the majority of modern batteries lack the liquid separators of the early voltaic cells, they function in much the same manner as the early 19th century devices, a tribute to the science of the early 19th century that first developed electric technology.

Selected Bibliography

Bordeau, Sanford P. *Volts to Hertz. . . . The rise of Electricity.* Minneapolis, MN: Burgess Publishing, 1982.
Karwatka, Dennis. *Technology's Past, Vol. 2: More Heroes of Invention and Innovation.* Ann Arbor, MI: Prakken Publications, 1999.
Krebs, Robert E. *Scientific Laws, Principles and Theories: A Reference Guide.* Westport, CT: Greenwood Press, 2001.
Spangenburg, Ray, and Diane K. Moser. *The History of Science in the Nineteenth Century.* New York: Facts on File, 1994.

Electric Light (ca. 1800–ca. 1898): The 19th century can be said to have been a time when physicists began to understand the properties of electricity and its relationship to magnetic force. More than any other scientific areas of study in the 19th century, the investigation into the properties of electricity would have the greatest influence on society and industry since the time of the Scientific Revolution. From the early studies of electromagnetic theory (see ELECTROMAGNETIC THEORY), to the defining of the mathematical laws of electrical potential (see ELECTRICITY), physicists during this time were developing a true appreciation for the potential of electrical forces. However, unlike the scientific discoveries of preceding centuries, in the later years of the 19th century inventors applied their scientific knowledge to the design of commercially profitable devices designed to improve people's standard of living. These inventors, many of whom did not completely understand the scientific principles of the items they were designing, formed an important link between science and technological advance. The invention of the incandescent electric lightbulb is an excellent example of this transfer between science and society.

Various forms of lighting have been in existence since the beginnings of civilization. However, before a usable electrical light could be invented scientists first had to develop a mechanism for delivering a sustained electrical current. Static electricity generators had been in existence since the 17th century, but it was the invention of the electric cell in 1800 by the Italian scientist Alessandro Volta that provided a source of constant current (see ELECTRIC CELL). Within a few years (ca. 1808) a number of investigators, most significantly the English scientist Humphry Davy, were experimenting with primitive forms of electric lights. These early instruments were designed to either generate a spark between two conductors or heat a small wire until it glowed. Over the next several decades both of these approaches were explored as mechanisms to deliver electric light. The invention of the electric generator in 1831 by the English physicist Michael Faraday also had an influence on the development of electric lights. The electric generator gave researchers the ability to continuously generate and supply an electrical current (see ELECTRIC POWER). Electric batteries were the preferred power source for the early electric lights, but over the next several years a number of improvements were made to electrical generators to increase their reliability and power output.

One of the earliest forms of electric lighting was the arc light. Although scientists had experimented with arc lights since Davy's experiments in the early years of the century, they did not become a reliable form of electric lighting until the work of Paul Jablochkoff (1847–1894) in the 1870s. Arc lights, called Jablochkoff's Candles, consisted of two parallel carbon rods mounted inside brass tubes. When power was supplied an electrical arc passed between the two carbon rods. Unfortunately, the rods had to be replaced each time the device was turned off. Nevertheless, Jablochkoff's Candles generated a tremendous interest in electric light and were frequently found along the streets of Paris and London by the late 1870s (see Figure 11). They also were unique in that they could be powered by both alternating (AC) and direct (DC) current, a useful characteristic since national standards for electric power had yet to be established (see ELECTRIC POWER). A number of improvements were made to arc lights, but a major drawback remained that the heat generated from the arcs degraded the carbon rods. This limited not only the widespread use of the lights, but also their distribution to the general public. The invention of the incandescent lamp (see later) marked the end of arc lighting. Still, it is interesting to note that various forms of arc lights persisted well into the 20th century and were used in various military campaigns until World War II.

The concept of using a glowing wire as a source of light began, like the arc light, with the early studies of electricity in the 19th century. Unfortunately, the technology did not exist for the development of a light that could produce enough output to compete with other forms of lighting, including arc lighting. The primary obstacle was that when the filament was heated to high enough temperatures to produce a usable glow, the filament itself frequently degraded, leaving a black coating inside the bulb. One solution was to produce a vacuum within the fixture, but a sufficiently powerful vacuum pump was not available until the mid-1860s. By 1881 four inventors had developed the technology to produce incandescent lamps. There were two English inventors, St. George Lane Fox and Joseph Swan (1828–1914), and two Americans, Thomas Edison (1847–1931) and Hiram Maxim (1840–1916). Of these,

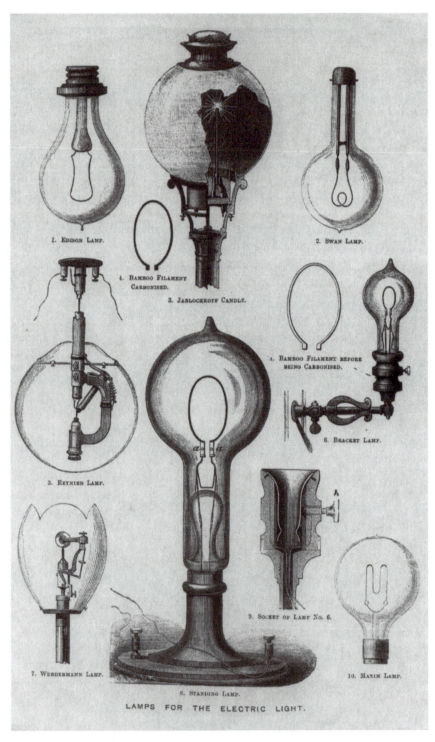

Figure 11. An example of electric lamps used in the 19th century. (Hulton Archive/Getty Images)

Edison and Swann are the most recognized for promoting and developing the modern forms of the incandescent lamp (see Figure 11).

The basic form of the incandescent lamp was similar for both the Edison and Swan inventions. At least initially, for both instruments a carbon filament was placed within a glass bulb, into which a vacuum was introduced. An electrical current was delivered to the filament via connections made of platinum, which was already recognized to conduct electricity better than most other materials. The Edison and Swan lights differed in the construction of the filament. Swan's light used a chemically treated paper as the filament, whereas in Edison's light the filament was a small piece of bamboo. Both inventors also proposed several other improvements to the light that increased the efficiency of the lamps in terms of the magnitude and quality of light produced and the operational length of the filament. The invention of electrical power stations and **alternating current** by Edison in the later years of the 18th century provided a means of delivering power (see ELECTRIC POWER), which served to increase the popularity of the electric incandescent light.

Incandescent lights were the more popular form of electric lighting at the end of the 19th century, but other forms of electric-powered lights were also being explored during this century. For some time prior to the invention of incandescent and arc lighting, scientists had recognized that an electric charge had the ability to cause a gas to glow. Faraday experimented with various forms of the phenomenon, but it was the work of the German scientist Johann Heinrich Geissler that produced the first light that was powered by charging a gas. Geissler designed a bulb that had a platinum wire passing through it. Earlier attempts had failed because the wire and bulb expanded unevenly when heated. By using platinum wire and the correct form of glass, Geissler was able to correct this problem. The result (ca. 1851) was the Geissler tube, an early form of electric light. At the end of the century Walther Nernst designed an electric light that used a cerium oxide filament to compensate for the damage caused by the high heat of the light. This highly efficient light was widely used in the early 20th century.

There can be little doubt that modern society has been transformed by the invention of the electric light. Although most modern lamps function in the same manner as the early lights invented by Edison and Swan, a number of improvements have been made to the technology since the 19th century. For example, in 1910 the carbon filaments were replaced by tungsten, a heat-resistant metal that greatly increased the efficiency and durability of the lightbulb. Later, inert gases such as argon and krypton were introduced to the interior of the bulb to prolong the life of the filament. There are many other modern forms of electric lights, including fluorescent, halogen, mercury vapor and sodium vapor lights. However, the basic technology of the electric light is a product of the 19th century scientists who first realized that electricity could be harvested for practical purposes.

Selected Bibliography

Bordeau, Sanford P. *Volts to Hertz. . . . The Rise of Electricity.* Minneapolis, MN: Burgess Publishing, 1982.

Bowers, Brian. *Lengthening the Day: A History of Lighting Technology.* New York: Oxford University Press, 1998. Karwatka, Dennis. *Technology's Past: America's Industrial Revolution and the People who Delivered the Goods.* Ann Arbor, MI: Prakken Publications, 1996.

Electric Power (1831–1893): The experimental scientific study of electricity began in the first years of the 17th century. However, the majority of these early studies were confined to examinations of static electricity. The invention of the electric cell in 1800 by Alessandro Volta allowed physicists to examine the principles of electrodynamics, or the movement of electricity between two points (see ELECTRIC CELL). In the first several decades following this invention scientists began a dedicated study of the electricity (see ELECTRICITY), which eventually led to the formation of a unified electromagnetic theory (see ELECTROMAGNETIC THEORY). One of the most important pioneers in the study of the electricity in the early 19th century was the English physicist Michael Faraday. Faraday lacked a formal education in the theory of physics, but he had a natural interest in, and ability to understand, electrical principles. Faraday is recognized as the pioneer in the understanding of electromagnetic fields and the process of electromagnetic induction (see ELECTRICITY, ELECTROMAGNETIC THEORY).

As part of his experimental approach to electricity, Faraday described the first true electrical generator in 1831. Static electricity generators had been in existence since the 17th century, but they lacked the ability to generate anything but a brief electrical charge. During his experiments Faraday had discovered that a magnetic field had the ability to generate an electrical current. He proposed that it was possible to generate a constant electrical current by moving wires through a magnetic field. Using a copper disk mounted between brass supports, Faraday was able to generate a direct current. Although Faraday had described and built the first electrical generator, he used it more to describe his principles of electromagnetic induction (see ELECTROMAGNETIC THEORY) than to generate current, probably because the electric cells designed by Volta earlier in the century provided sufficient current for the majority of his studies. Still, others became interested in the idea and within just a few years a number of variations of Faraday's machine were being produced, including one of the first in which the coils were placed in motion around the magnets.

The industrial world slowly adapted to the use of electrical generators for much the same reason as Faraday and others did not use them for their research—the electrical batteries of the time delivered a greater current. However, by the 1840s a number of heavy industries were developing that had need of greater amounts of electrical current. The first application of Faraday's electrical generator was in 1844 by the English businessman John Woolrich, whose industry was involved in the process of electroplating. The inventors of electric lighting, specifically a form of lighting called arc lighting, also found that electric generators were useful (see ELECTRIC LIGHT). Like the electroplating industry, lighthouses required current levels that quickly depleted batteries, and thus by the 1850s some lighthouses were being refitted with steam-driven electrical generators.

Several improvements were made to Faraday's design over the next two decades. The majority of the new designs worked to improve the quality of the magnetic field and thereby generate a higher level of current. An added benefit of the new generators was that they required less space and thus could be used by a greater variety of industries. One of these improvements used a portion of the generator's own output current to internally power the magnetic coils, a process called self-excitation. This is the basis for the modern form of electrical generators called dynamos (see Figure 12). By the 1870s inventors had made generators that were only a fraction of the size of early generators and produced significantly more power.

For some time scientists and industrialists had recognized the usefulness of a machine that could convert electrical energy to mechanical energy. Based on the reciprocating design of steam engines (see STEAM ENGINES), Joseph Henry built a prototype of an electric motor in 1831. The design had been suggested earlier in the century by both Faraday and Peter Barlow (1776–1862). However, Henry's machine was the first to be designed that caught the attention of industry. This machine simply used magnetic forces to drive an armature. The primary drawback of the early electric motors was that they were competing with the more powerful steam engines, which did not require an external source of electricity for operation. However, some inventors, such as Thomas Davenport (1802–1851) were able improve the generators and make them attractive for industries that did not require heavy current levels. By the mid-1830s Davenport's electrical motors were powering printing presses and other small industries.

Early generators and motors operated using direct current, which limited the amount of distance that the electricity could be transmitted. This limita-

Figure 12. An example of a 19th century electrical power generator called a dynamo. (Hulton Archive/Getty Images)

tion was due to the principles of electricity, namely Ohm's law (see ELEC-TRICITY). The low voltage output of the electrical generators required that the electricity travel to the site at a high current, which in turn required expensive electrical conductors. By 1882 the American inventor Thomas Edison had developed central power stations in New York City (see Figure 13) to deliver power for his incandescent lamps (see ELECTRIC LIGHT), but they were still limited in the distance that they could transmit power. Power generation thus required a number of small generating stations serving local customers. To make the use of electrical motors more attractive the ability to transfer electrical power long distances from a distant power source to the consumer was needed. A solution was to use alternating current (AC). Alternating current could be transmitted over long distances at high voltage and low current, lessening the need for bulky conductors. Unfortunately, the majority of electric motors used direct current, which limited the marketability of AC current. However, in 1888 Nikola Tesla (1856–1943) invented the first practical AC motor that made reliable use of alternating current, also

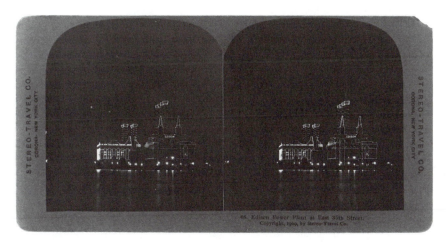

Figure 13. A photo of the Edison power-generating plant in New York City taken in the early 20th century. (Library of Congress photo collection)

called a polyphase induction motor. A fierce competition began between the direct current suppliers, largely under the control of Edison's companies, and investors of alternating current to provide local power. However, Tesla's invention gave alternating current the advantage. It was now possible to generate and transmit alternating current over long distances. The 1893 opening of an alternating current generation plant at Niagara Falls, New York, and its ability to deliver inexpensive electrical power over long distances, sealed the future of alternating current.

It is interesting to note that modern society is still paying the price for the battles between the supporters of alternating and direct currents. Because direct current can't be transmitted over long distances, some people suggested building a canal from Niagara Falls to provide power to industries away from the natural water source. However, with the invention of alternating current, these canals were no longer needed. Since that waterway was never completed, it was converted into a waste disposal site called Love Canal.

There were other advances in the study of electric power. Earlier in the century, Faraday's study of electromagnetic induction had shown that the electric current in a wire had the ability to generate a current in a parallel wire by the process of electromagnetic induction (see ELECTROMAGNETIC THEORY). This principle is the basis for the operation of a transformer, and in fact Faraday's coil is considered to be one of the first transformers invented. A precursor to the transformer was a device called the induction coil. Rather than a steady current, the induction coil produced brief high-voltage sparks. Early versions of the device were used in medical studies of electricity and to interrupt the constant current of early batteries (see ELECTRIC CELL) to study the properties of static electricity. However, researchers were beginning to notice that the coils could serve another function. By changing the number of coils in the opposing wires, and the spacing between the wires, it is possible to change the voltage and current being generated across the coils. The American scientist Joseph Henry (1797–1878) was one of the first to recognize this, and this knowledge played an important role in his eventual development of powerful electromagnets (see ELECTRO-MAGNETIC THEORY). Transformers became an important component of the power industry. AC current could be transferred over long distances, but the high-voltage at which it was transmitted was not usable by most electrical devices. In 1885 William Stanley (1858–1916) invented a transformer that converted the transmitted AC current to a low-voltage, high-current state. This is frequently called a step-down transformer, although other versions exist that raise the voltage. Almost all electronic devices contain a power transformer to reduce the electrical supply to a level that is usable by electronic circuits. The power delivered for use in North American is different from that used in the remainder of the world. Power in the United States is transmitted at 60 hertz (see ELECTRICITY) and 120 volts, but in Europe the rate of transmission is 50 hertz at 220 volts.

The availability of electric power enabled the creation of other scientific and technical inventions in the 19th century. Perhaps the greatest of these was the electric light bulb (see ELECTRIC LIGHT). Other devices, such as the telegraph and electromagnets, which had previously been powered by large batteries, were quickly connected to electrical generators to give them a more reliable power source (see ELECTRICITY, TELEGRAPH). At the end of the century large-scale power generation became possible with the invention of the steam turbine (see STEAM ENGINES) in 1884 by Charles Parsons (1854–1931). Electric power is one of the most significant technological advances in human history. It represents the perfect link between science, technology and society. The scientific community entered the 19th century without a clear understanding of the principles of electricity. Through dedicated experimentation, physicists were able to not only develop a clearer picture of electricity, but also in the process develop the capabilities for electrical power. The availability of electrical power forever changed the structure of society. For example, electrical communication systems such as the telegraph and radio linked remote areas of the planet to the centers of civilization, while electrical-powered industry produced consumer goods at a rate higher than that of any other time in history. The availability of electrical power to the average person greatly enhanced the quality of life. Our modern electrical society is the direct result of the work 19th century pioneers such as Faraday and Edison.

Selected Bibliography

Bordeau, Sanford P. *Volts to Hertz. . . . The rise of Electricity.* Minneapolis, MN: Burgess Publishing, 1982.
Bud, Robert, and Deborah Jean Warner (eds.). *Instruments of Science.* New York: Garland Publishing, 1998.
Krebs, Robert E. *Scientific Laws, Principles and Theories: A Reference Guide.* Westport, CT: Greenwood Press, 2001.
Meyer, Herbert W. *A History of Electricity and Magnetism.* Cambridge, MA: MIT Press,1971.

Electricity (1800–1879): The spectacle of a summer evening thunderstorm is a testimony to the power of electricity as a force of nature. the relationship between lightning and electricity was not recognized until the work of the American scientist and statesman Benjamin Franklin in the 18th century, although scientists a century earlier had initiated the experimental investigation of electrical forces. In the early years of the 17th century the research of the English scientist William Gilbert (1544–1603) helped to define some of the differences between electricity and magnetism. Charles Augustin Coulomb (1736–1806) provided a more formal separation of these forces later in the 18th century. In the mid-17th century the German scientist Otto von Guericke invented a device that could generate static electricity and, in the process, effectively began the experimental study of electrical forces. However, until the 19th century scientists were primarily confined to the study of static electricity only, since they lacked the ability to generate a constant electrical current for prolonged studies. Still, some progress was made during the 18th century, such as Franklin's studies of atmospheric electricity and his (and others) realization that electricity may have two forms, which are recognized as positive and negative charges. Many considered that electricity behaved as a fluid, and theories were debated that electricity acted as a single or double fluid moving between two points. As with similar studies of light and heat (see HEAT, LIGHT), the 19th century investigations of electricity would link theoretical considerations with experimental evidence and mathematical proofs to provide a significantly clearer picture of the properties of this force.

Perhaps the most important invention for the study of electricity was made in 1800, when the Italian scientist Alessandro Volta designed the first electric cell, sometimes called a voltaic cell or voltaic pile (see ELECTRIC CELL). Unlike previous instruments, this device allowed physicists to generate and store an electrical current. Essentially a battery, the electric cell gave physicists the ability to deliver a constant electrical charge for experimental studies. This effectively began the study of electrodynamics, or the movement of electricity, and initiated a century of study into the properties of electricity. As scientists began to pass electrical currents through coils of wire, they noticed a definite relationship between electricity and magnetic lines of force (see ELECTROMAGNETIC THEORY). The first experimental evidence for this relationship was provided by Hans Christian Oersted (1777–1851) in 1819, He demonstrated that an electrical current had the ability to deflect the needle of a nearby compass. The German physicist Johann Schweigger (1779–1857) used Oersted's discovery to construct the first galvanometer, an instrument that measures electrical current. However, it was André-Marie Ampère (1775–1836), a French scientist, who provided a more direct connection between electricity and magnetism. Ampère's experiments (see ELECTROMAGNETIC THEORY) demonstrated that magnetic lines of force were generated from electrical currents. Michael Faraday experimentally proved the reverse idea, that an electrical current could be generated by a magnetic field, around 1821. Faraday not only supplied the remaining link between the forces of magnetism and electricity, he also suggested that magnetism exists as fields of force radiating outward from the source (see ELECTROMAGNETIC THEORY). This led to a final synthesis of a unified electromagnetic theory (see ELECTROMAGNETIC THEORY) in 1873 by the Scottish physicist James Clerk Maxwell (1831–1879).

Ampère's and Faraday's experiments initiated an intense interest in the study of electricity, which in turn led to a number of important advances. In 1823, an English physicist named William Sturgeon (1783–1850) invented the electromagnet by placing an iron bar inside a helix of wire (called a solenoid), When an electrical current was applied to the helix, the resulting electromagnetic field had the ability to pick up an item several times the weight of the magnet. By 1831 Joseph Henry, an American physicist, had invented electromagnets capable of lifting more than a ton of iron. The electromagnet quickly found application in the developing heavy industries of the 19th century (see STEEL). Henry was also responsible for the design of electromagnetic motors and the telegraph (see ELECTRIC POWER, TELEGRAPH). Others, such as Georg Ohm (1789–1854), began to mathematically define the movement of electricity. Ohm recognized that electricity might flow in the same manner as heat (see HEAT). By using wires of different lengths and thickness, Ohm was able to vary the resistance of the wires. By 1827 he was able to demonstrate that the flow of electricity was inversely proportional to the resistance of the wires. This is represented by the formula:

$$iv/r$$

where i represents the current, v the voltage and r the resistance of the circuit. In other words, the higher the resistance, the lower will be the current. Ohm's law plays an important role in the study of conductivity and the design of all electrical circuits. Around the same time another scientist, Thomas Seebeck (1770–1831) began to recognize that the flow of an electrical current could also be regulated by heat. This is called the Seebeck effect and marked the beginning of the study of thermoelectricity. Although this area did not significantly influence the direction of 19th century science, it would play an important role in the design of superconductors in the late 20th century.

In addition, both Ampère and Faraday made contributions to the study of electricity other than those that led to the electromagnetic theory. Ampère is considered to have developed the first true galvanometer for the measurement of current. Although previous instruments had the ability to detect electrical currents, Ampère's machine allowed researchers to record not only the strength of the current but also the direction in which it was moving. In addition, Ampère invented one of the first solenoids, which are long coils of wire used to enhance the strength of an electromagnetic field. He also suggested that magnetism and electricity might be utilized for long-distance communications (see TELEGRAPH). Faraday's work on the electromagnetic induction of current from one coil or wire to another became the basis for an electrical device called a transformer, an important component of most modern electrical systems. Furthermore, Faraday is recognized as the inventor of the first practical electrical generator (1831). This instrument converted the motion of a copper wheel between two coils into a constant electrical potential (see Figure 14). The wheel could be powered by any number of different sources (water, fuels, etc.). The initial purpose of this machine was to supply a constant current source for Faraday's experiments, although it quickly became evident that the machine had a number of commercial applications (see ELECTRIC POWER).

As noted, many of these electrical discoveries were quickly converted to commercial use. The invention of reliable electrical generators made it possible to deliver electrical power over a distance, which resulted in the widespread use of other electrical devices. For example, in 1879 the American inventor Thomas Edison designed the first incandescent electric light (see ELECTRIC POWER). Although the telegraph had been in existence since the early 1830s, the invention of the electrical relay by Henry (ca. 1835) allowed telegraph wires to span long distances (see TELEGRAPH). Another electrical component, the transformer, was deigned in 1885 by William Stanley. This device could vary voltage and current from high to low. Heinrich Rudolph Hertz's (1857–1894) study of electromagnetic waves

Figure 14. One of the first electrical generators of the 19ᵗʰ century, designed by Michael Faraday in 1831. (Hulton Archive/Getty Images)

(see ELECTROMAGNETIC THEORY) in 1888 provided useful information the invention of the radio, a wireless communication device (see RADIO).

The importance of these investigators to the study of electricity is easily illustrated by examining the names assigned to electrical units. The *ohm* represents a unit of resistance, and the *ampere* is associated with the measure of current. The term *volt* represents the difference in electrical potential between two points. The term *henry* is a unit of measurement for the electromagnetic field generated by an electrical current in a wire. In addition, the terms **joules** and **watts** are frequently associated with electricity, although they describe energy and power and originated primarily with the development of improved steam engines (see ELECTRIC POWER, STEAM ENGINES). The mathematical values of these units were first standardized by the German scientists Carl Gauss (1777–1855) and Wilhelm Weber (1804–1891) beginning in the 1830s. Although the values of these units have been refined since that time, the need for comparing the measurements being obtained from various labs was recognized early in the study of the electricity. Electricity plays an important role in the structure of modern society, and most people in the Western world take it for granted. However, when compared to other areas of science, the study of electricity is a relatively recent event. The fact that it has so recently become such a dominant factor in our culture is a reflection of the importance of these 19th century studies and a tribute to the early electrical pioneers.

Selected Bibliography

Bordeau, Sanford P. *Volts to Hertz. . . . The rise of Electricity*. Minneapolis, MN: Burgess Publishing, 1982.
Krebs, Robert E. *Scientific Laws, Principles and Theories: A Reference Guide*. Westport, CT: Greenwood Press, 2001.
Meyer, Herbert, W. *A History of Electricity and Magnetism*. Cambridge, MA: MIT Press, 1971.
Spangenburg, Ray, and Diane K. Moser. *The History of Science in the Nineteenth Century*. New York: Facts on File, 1994.
Williams, L. Pearce. "André-Marie Ampère." *Scientific American* 260(1): 90–97.

Electromagnetic Spectrum (1800–1896): The question of whether light exists as a wave or a particle occupied the efforts of physicists from the time of the Scientific Revolution until the 19th century. However, by the second decade of the 19th century scientists had demonstrated that the majority of the properties of light, including **diffraction** and **refraction**, could be explained with light as a wave. Much like sound waves, waves of light can be measured by the distance of their wavelength, or the distance between two consecutive troughs or peaks of the wave (see LIGHT for figure). The discovery that light has a complex structure predates the 19th century. Almost two centuries earlier Isaac Newton used a prism to demonstrate that white light consisted of a spectrum of individual colors and suggested that these colors may have different properties. This spectrum is called the electromagnetic spectrum. By the late 19th century scientists had begun to define the relationship between electricity, magnetism and light and in the process developed the electromagnetic theory (see ELECTROMAGNETIC THEORY). The study of the electromagnetic spectrum in the 19th century followed two distinct paths. First, a group of physicists began to recognize that portions of the electromagnetic spectrum existed outside those wavelengths visible to the human eye. Second, chemists and physicists began to study the wavelengths of the spectrum emitted or reflected by an object using a new instrument called a spectroscope. The use of the spectroscope began the study of spectral analysis (see SPECTROSCOPE). Both of these areas of investigation had a significant influence on the science of the times.

In 1800 William Herschel, an English astronomer, conducted an experiment on the thermal capabilities of light. It was well known that visible light produced heat, but Herschel was interested in the heat-producing abilities of the various regions of the visible spectrum. He noticed that as a thermometer was moved along the visible spectrum the red regions registered the highest temperatures. However, as he moved the thermometer past the red end of the spectrum, and out of the range of visible light, the temperature continued to rise. He did not notice a similar increase beyond the violet end of the spectrum. Herschel used the term *infrared* to refer to the region beyond the red end of the spectrum, and it became the first part of the electromagnetic spectrum to be detected outside the visible wavelengths. The following year Wilhelm Ritter (1776–1810), a German physicist, discovered that there was a form of invisible light beyond the violet end of the spectrum, although it did not have the heat properties of the infrared region. Ritter knew that the chemical compound silver chloride turns dark when exposed to light. When Ritter placed a plate containing silver chloride in the area past the violet end of the spectrum, it also darkened. Ritter had identified the *ultraviolet* portion of the electromagnetic spectrum and in the process extended the electromagnetic spectrum in the opposite direction of Herschel's discovery. Ritter's discovery of the breakdown of silver chloride would eventually find application in the invention of photography later in the century (see PHOTOGRAPHY).

Both Hershel and Ritter had demonstrated a form of light that was not visible to the human eye, but in the early years of the 19th century the scientific community was still attempting to understand the basic properties of light. At the start of the century the majority of people in the scientific community were still following Isaac Newton's 17th century thinking that light was a particle. However, by 1818 the work of scientists such as Thomas Young and Augustin Fresnel clearly demonstrated that light existed as a wave (see LIGHT). This led to the recognition that magnetism, electricity and light are all interconnected. The Scottish physicist James Clerk Maxwell is credited with the first presentation of this relationship as a unified theory in 1873, although a number of other important scientists made contributions to his idea (see ELECTROMAGNETIC THEORY). These discoveries played an important role in the study of the electromagnetic spectrum later in the century.

The next advance in the study of the electromagnetic spectrum was made in 1888 by the German physicist Heinrich Rudolf Hertz. Hertz was the first to actually measure a wavelength outside the visible spectrum. Hertz generated the electromagnetic waves for his experiment using a high-voltage alternating current (see ELECTRICITY). It was possible to generate an electromagnetic wave by passing this current between two metal spheres. To detect the wave, Hertz designed a simple receiver consisting of a single loop of wire with a single gap. He realized that when the high-voltage current arced between the metal spheres, the resulting electromagnetic wave should move across the room and generate a spark in the receiver. By moving the receiver around the inside the room Hertz was able to determine not only the shape of the waves but also the wavelength. He calculated that this invisible radiation had a wavelength of about 66 centimeters (24 inches). Not only was this experiment important in the study of the electromagnetic spectrum, but it also provided an important verification of Maxwell's theories. The form of radiation that Hertz studied are now recognized to be radio waves (see Figure 15). Six years later the Italian scientist and inventor

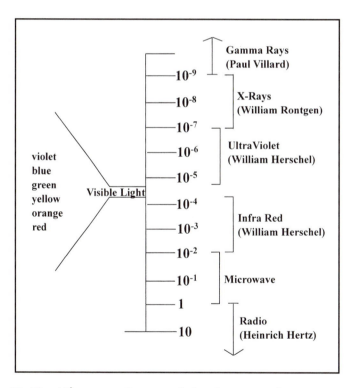

Figure 15. The 19th century discovery of the electromagnetic spectrum, with the principle discoverers. The numbers indicate the wavelength in centimeters (cm). Microwaves were discovered during the 20th century.

Guglielmo Marconi (1874–1937) was successful in using this form of electromagnetic radiation to invent the first form of wireless communication (see RADIO).

In 1895 the German physicist William Röntgen (1845–1923) was studying the fluorescent capabilities of the recently invented cathode-ray tube (see CATHODE-RAY TUBE). In one of his experiment Röntgen placed a dark paper coated with the chemical compound barium platinocynide around the cathode-ray tube (CRT). Röntgen noticed that when the CRT was turned on, the coated paper glowed with a faint light that did not originate within the CRT. The glow continued even when the paper was placed some distance from the original source. Röntgen theorized that the CRT device must be emitting some form of previously undiscovered radiation, which he called *x-rays* (see Figure 15). X-rays would become a popular area of study in the early 20th century and resulted in a revolution of the study of the properties of the electromagnetic spectrum. Röntgen received the first Nobel Prize in physics in 1901 for his work. X-rays now serve as an extremely important tool not only for physicists, but also for astronomical studies, the biochemical examination of molecular structure and the practice of medicine. Their discovery probably represents one of the most important advances in the study of the electromagnetic spectrum in the 19th century.

The discovery of x-rays resulted in a renewed interest in the study of the electromagnetic spectrum. In 1896 the French physicist Antoine-Henri Becquerel (1852–1908) was continuing Röntgen's study of using fluorescent materials to generate x-rays. In his experiments he used a crystal made from potassium uranyl sulfate, which was known to fluoresce when exposed to sunlight. He used photographic plates to detect the images because any x-rays that were generated would create a ghostlike image on the plate. However, one day Becquerel left a piece of crystal on a covered photo plate in a closed drawer. When he later developed the film he noticed that an image had been produced from the crystal without it being exposed to sunlight. Thus, he concluded that some compounds emit particles without being directly stimulated. This was an exceptionally important discovery and resulted in Becquerel being awarded the Nobel Prize in physics in 1903. These particles were further classified and named *alpha* and *beta particles* by the English physicist Ernest Rutherford (1871–1937) in 1896.

The last major 19th century discovery involving the electromagnetic spectrum was that of *gamma rays*. In 1900 the French physicist Paul Villard (1860–1934) discovered that the radiation being emitted by uranium consisted not only of the previously identified alpha and beta waves but also of a radiation that possessed a very short wavelength and extremely high intensity (see Figure 15).

The study of the electromagnetic spectrum spanned the 19th century. Science entered the century without a clear understanding of the visible spectrum of light and in considerable debate on the very nature of light. Over the course of the 19th century scientists gradually began to explore the boundaries of the electromagnetic spectrum and discovered that a vast array of radiation existed. With the exception of **microwaves**, the majority of the electromagnetic spectrum was discovered in the 19th century. This study had a significant influence on the science of the 20th and 21st centuries. The study of the electromagnetic spectrum was influential in the study of radioactivity in the early 20th century, the invention of the radio and the beginning of x-ray studies of molecules. Modern society utilizes virtually the entire electromagnetic spectrum, a reflection of the importance of these 19th century studies.

Selected Bibliography

Bordeau, Sanford P. *Volts to Hertz. . . . The rise of Electricity*. Minneapolis, MN: Burgess Publishing, 1982.

Karwatka, Dennis. *Technology's Past, Vol. 2: More Heroes of Invention and Innovation*. Ann Arbor, MI: Prakken Publications, 1999.

Park, David. *The Fire within the Eye: A Historical Essay on the Nature and Meaning of Light*. Princeton, NJ: Princeton University Press, 1997.

Spangenburg, Ray, and Diane K. Moser. *The History of Science in the Nineteenth Century*. New York: Facts on File, 1994.

Electromagnetic Theory (1819–1888): Starting with the Scientific Revolution of the 17th century, scientists were beginning to experimentally investigate the natural world, including the physical forces of electricity and magnetism. Early in these studies many of the investigators frequently confused the attractive forces of electricity and magnetism. William Gilbert, the English pioneer in the modern study of magnetism and one of the first proponents of the use of experimentation to support scientific theories, was also the first to make a distinction between the physical properties of electricity and magnetism. Gilbert understood that magnetism was due to the presence of magnetic iron ore in a body, while an electrical attraction could occur between almost any objects. However, later studies of electricity and magnetism in the 17th and 18th centuries frequently overlooked that these forces were interconnected. This was primarily because of the lack of an instrument that could sustain electrical currents for experimental study. However, by the 19th century advances in the design of electrical storage devices, later called cells and batteries, was enabling the study of electricity (see ELECTRIC CELL). At the same time, physicists were initiating a more detailed examination of much of the natural world. In their investigations of light they began to recognize that light functioned as a wave and that visible light was simply a small component of the electromagnetic spectrum (see ELECTROMAGNETIC SPECTRUM, LIGHT). By the middle decades of the century scientists were also making a dedicated effort to experimentally explore the nature of electricity (see ELECTRICITY). It was these studies that established the experimental proof necessary for the development of a unifying theory on the relationship of electricity and magnetism.

The first evidence of a link between electricity and magnetism began with the experiments of Hans Christian Oersted around 1819. Oersted, an instructor of physics at the University of Copenhagen, unintentionally demonstrated to his students that an electrical current passing through a wire had the ability to deflect the needle of a nearby magnetic compass. Furthermore, it was later demonstrated that the magnetic force was circular, meaning that it was projected in all directions around the wire. Within a few years Jean-Baptiste Biot (1774–1862), and others, proved that the magnetic force decreased with the distance from the wire according to the inverse-square law. The inverse-square law was well recognized by physicists, having previously been applied to earlier studies of both magnetism and gravitation, and this finding provided an important verification that this was a natural force.

Oersted's work had a tremendous influence on a number of other researchers of electrical force. The most important of these was the French scientist André-Marie Ampère. Ampère made a number of exceptionally important contributions to the 19th century study of electricity, especially in the field of electrodynamics, or the movement of electricity between two points (see ELECTRICITY). While others, such as Charles Augustin Coulomb were defining the similarities of the physical laws of electricity and magnetism (see ELECTRICITY), it was Ampère's experiments with the generation of magnetic fields by electrical currents that provided the crucial evidence that the two forces were interconnected. Ampère noted that when an electrical current is delivered in the same direction down two parallel wires the generated magnetic fields attract the wires. When the current was passed through the wires in opposite directions the wires were repelled. Furthermore, Ampère proposed that the magnets were a collection of electrical currents, and not a fluid compound as many of his time thought. This idea had a tremendous impact on the later development of a unified electromagnetic theory.

Michael Faraday, a common Englishman with an intense interest in electricity despite a lack of formalized training in physics, is credited with unifying the study of electricity and magnetism. In 1821 Faraday provided some of the most conclusive experimental evidence of electromagnetism. Based on the experiments of Oersted and Ampère, Faraday decided to determine whether a magnetic field had the ability to generate an electrical current. Faraday's experiment, now considered a classic example of experimental design, used an instrument called a galvanometer (see ELECTRICITY) to measure the flow of current. First, Faraday connected a coil of iron wire to the galvanometer. Then he attached an electrical battery (see ELECTRIC CELL) to a second coil of wire (see Figure 16). Oersted had previously demonstrated that current could be used to generate a magnetic field, and Ampère had shown that the magnetic fields of two wires interacted. In his experiment Faraday noticed that when an electrical current was applied from the battery to one coil of wire, a brief electrical current was detected by the galvanometer on the second wire coil. Faraday had demonstrated the process of *electromagnetic induction* and in the process had laid the groundwork for the subsequent development of electrical motors and generators (see ELECTRIC POWER). What further intrigued Faraday was that although the current from the battery was constant the galvanometer in the second coil did not register a constant flow of coil. A brief spike of current was detected by the galvanometer each time the battery was cycled on and off. To Faraday, this suggested that the magnetism of the coil was to the result of lines of force being projected from the wire and it was the initial interaction of these lines of force that induced the current in the second coil. Figure 17 is a photograph of one of Faraday's electromagnets. Faraday presented his ideas to the Royal Society in 1831 and in the process revolutionized the study of physics. Faraday did

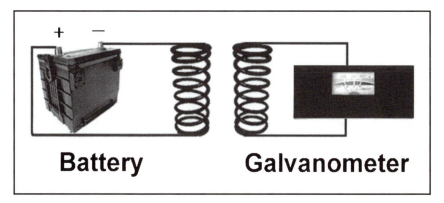

Figure 16. Faraday's experiment to determine whether a magnetic field has the ability to generate an electrical current. The galvanometer is used to detect current, and the battery provides the power for the two coils in the center.

Figure 17. A photograph of one of the electromagnets used by Faraday for the study of the electromagnetic theory. (Hulton Archive/Getty Images)

not limit his studies to the interaction of electricity and magnetism. He also demonstrated (1845) that an electromagnetic field has the ability to change the polarization of light, a phenomenon frequently called the Faraday effect. These studies helped demonstrate that light, magnetism and electricity were all connected, as is evident in the study of the electromagnetic spectrum (see ELECTROMAGNETIC SPECTRUM).

Without formal training, Faraday lacked the mathematical ability to explain the process of electromagnetism. However, his experimental evidence was so strong that soon several powerful physicists, such as James Clerk Maxwell and William Thomson, were attempting to establish the underlying principles of electromagnetism. Maxwell, based on the work of Thomson and others, developed a series of elaborate mathematical equations that accounted for the previously observed properties of electromagnetism. His calculations demonstrated that electricity and magnetism coexisted and that the two were interchangeable. These calculations, presented in 1873 as Maxwell's *Treatise on Electricity and Magnetism*, are considered to be the foundation of modern electromagnetic theory. Furthermore, his calculations indicated that light may also be a form of electromagnetic radiation, because the velocity of electromagnetic lines of force equaled that of the velocity of light. In 1888 Heinrich Rudolf Hertz provided the experimental proof of Maxwell's theory. As part of his study of electromagnetic waves (see ELECTROMAGNETIC SPECTRUM), Hertz experimentally measured the wavelength of an electromagnetic wave generated from an electrical source (see RADIO).

The experimentation behind the development of the electromagnetic theory represents one of the more significant advances of the 19th century. Not only did it result in a scientific theory that unified the properties of electricity, magnetism and light, but it also resulted in some important technological advances. For example, the discovery of electrical induction gave rise to the invention of electrical generators, and the electromagnetic waves discovered in Hertz's experiments became the basis of wireless communication, or radio. From a purely scientific perspective, the electromagnetic theory is considered to be as important as Isaac Newton's universal theory of gravitation was in the 17th century. The electromagnetic theory revolutionized the study of physics and played a crucial role in the development of theoretical physics in the 20th century.

Selected Bibliography

Karwatka, Dennis. *Technology's Past, Vol. 2: More Heroes of Invention and Innovation*. Ann Arbor, MI: Prakken Publications, 1999.

Krebs, Robert E. *Scientific Laws, Principles and Theories: A Reference Guide*. Westport, CT: Greenwood Press, 2001.

Nye, Mary Jo. *Before Big Science: The Pursuit of Modern Chemistry and Physics 1800–1940*. New York: Twayne Publishers, 1996.

Purrington, Robert D. *Physics in the Nineteenth Century*. New Brunswick, NJ: Rutgers University Press, 1997.

Spangenburg, Ray, and Diane K. Moser. *The History of Science in the Nineteenth Century*. New York: Facts on File, 1994.

Elements (1801–1898): Modern chemistry recognizes an element as a substance that may not be broken down into smaller components by normal chemical reactions. However, the classification of elements has undergone substantial revision throughout the history of science. Prior to the Scientific Revolution, most scientists supported the ancient Greek philosopher

Aristotle's (384–322 B.C.E) Four Element theory. Although portions of this theory predate Aristotle by centuries, the theory contended that four prime elements (air, fire, water and earth) made up all matter. In the two centuries leading up to the 1800s chemists worked to redefine the physical properties of elements. In the 17th century the work of the English chemist Robert Boyle introduced the modern definition of an element and demonstrated that the properties of elements could be determined by experimental means. In the early 18th century the French chemist Antoine Lavoisier's studies indicated that the properties of compounds were dependent on the elements that they contained. In the early 19th century John Dalton, an English chemist, developed the foundations of the modern atomic theory (see ATOMIC THEORY). Despite the theoretical progress, relatively few new elements were discovered during these times. Isolation and classification of the majority of the 92 naturally occurring elements would have to wait for technological advances in the science of chemistry.

The discovery of the elements in the 19th century can be directly linked to improvements in both the experimental method and chemical instrumentation. Although more elements were discovered in the 19th century than during any other previous period of history (see Table 2), only a few of these elements had a direct impact on the science of the times. The first significant discoveries were sodium and potassium. New elements had regularly been discovered throughout the 18th century, most of which were metallic elements. However, by the end of the 18th century additional discoveries were hindered by the fact that a number of the metallic elements form close associations with oxygen, resulting in what are called metal oxides. This association made isolation of the individual elements within the compound difficult. However, with the invention of the battery in 1800 by Alessandro Volta (see ELECTRIC CELL) it had become possible to generate and sustain an electrical current. In 1807 the English chemist Humphrey Davy was experimenting with passing an electrical current through potash (potassium carbonate). He noticed that when potash was exposed to a strong electrical current, a small flame appeared at the cathode of the battery. In the same vicinity, small globules of metal began to form. These metallic globules reacted with the water to release hydrogen, a flammable gas. Davy named this element potassium. Shortly thereafter Davy performed the same experiment with sodium carbonate to generate elemental sodium. At that time the introduction of this experimental process, called electrochemistry, was more important to science than the elements themselves, although sodium and potassium are now recognized as essential elements to biological systems. Using this procedure, in 1808 Davy isolated the elements calcium, magnesium, strontium and barium from previously identified metal oxides of these elements. Over the

Table 2
List of Elements Discovered in the 19ᵗʰ Century. Symbol represents the atomic symbol of the element on the modern periodic table of the elements.

Element	Symbol	Year		Discoverer	Nationality
Niobium	Nb	1801		C. Hatchett	English
Tantalum	Ta	1802		A. Ekeberg	Swedish
Cerium	Ce	1803		J. Berzelius	Swedish
			and	W. Hisinger	Swedish
Iridium	Ir	1803		S. Tennant	English
Osmium	Os	1803		S. Tennant	English
Palladium	Pd	1803		W. Wollaston	English
Rhodium	Rh	1803		W. Wollaston	English
Potassium	K	1807		H. Davy	English
Sodium	Na	1807		H. Davy	English
Barium	Ba	1808		H. Davy	English
Calcium	Ca	1808		H. Davy	English
Magnesium	Mg	1808		H. Davy	English
Strontium	Sr	1808		H. Davy	English
Chlorine	Cl	1810		H. Davy	English
Iodine	I	1812		B. Courtois	French
Cadmium	Cd	1817		F. Strohmeyer	German
Lithium	Li	1817		J. Arfwedson	Swedish
Selenium	Se	1817		J. Berzelius	Swedish
Silicon	Si	1824		J. Berzelius	Swedish
Zirconium	Zr	1824[1]		J. Berzelius	Swedish
Aluminum	Al	1825		H. Oersted	Danish
Bromine	Br	1826		A. Balard	French
Thorium	Th	1828		J. Berzelius	Swedish
Vanadium	V	1830		N. Selfstrom	Swedish
Lanthanum	La	1839		C. Mosander	Swedish
Erbium	Er	1843		C. Mosander	Swedish
Terbium	Tb	1843		C. Mosander	Swedish
Ruthenium	Ru	1844		K. Klaus[2]	Russian
Cesium	Cs	1860		G. Kirchhoff	German
			and	R. Bunsen	German
Rubidium	Rb	1861		G. Kirchhoff	German
			and	R. Bunsen	German
Thallium	Tl	1861		W. Crookes	English
Indium	In	1863		F. Reich	German
			and	T. Richter	German
Helium	He	1868		J. Janssen	French
			and	J. Lockyer	English
Gallium	Ga	1874		P. Boisbaudran	French
Ytterbium	Yb	1878		J. C. G. de Marginac[2]	French

Table 2
Continued

Element	Symbol	Year		Discoverer	Nationality
Holmium	Ho	1879		P. Cleve	Swedish
Scandium	Sc	1879		L. Nilson	Swedish
Samarium	Sm	1879		P. Boisbaudran	French
Thulium	Tm	1879		P. Cleve	Swedish
Gadolinium	Gd	1880		J. C. G. de Marginac	French
Neodymium	Nd	1885		C. A. B. von Welsbach	Austrian
Praseodymium	Pr	1885		C. A. B. von Welsbach	Austrian
Dysprosium	Dy	1886		P. Boisbaudran	French
Fluorine	F	1886[1]		H. Moissan	French
Germanium	Ge	1886		C. Winkler	German
Argon	Ar	1894		J. Rayleigh	English
			and	W. Ramsey	English
Krypton	Kr	1898		W. Ramsey	English
			and	M. Travers	English
Neon	Ne	1898		W. Ramsey	English
			and	M. Travers	English
Polonium	Po	1898		M. Curie	Polish
Radium	Ra	1898		M. Curie	Polish
Xenon	Xe	1898		W. Ramsey	English
			and	M. Travers	English
Actinium	Ac	1899		A. Debierne	French

Notes:
1. Discovered prior to the 19th century. The entry indicates the first person to isolate the element.
2. Discovery is frequently credited to multiple investigators.

next few decades a large number of elements were identified using electro-chemistry (see Table 2).

As the number of identified elements increased, a need arose for a standardized system of nomenclature. One of the first attempts to organize the elements started around 1813 with the work of the Swedish chemist Jöns Berzelius. Berzelius was involved in the discovery of several elements, including cerium and selenium (see Table 2) and was also responsible for several important contributions to the study of atomic theory (see ATOMIC THEORY). One of his contributions to the study of chemistry was the introduction of the nomenclature system for the elements that persists to this day. Before Berzelius, chemists primarily relied on a system developed by Dalton that utilized spherical line drawings to represent the chemical elements.

Berzelius instead suggested that the elements be identified using the first letter of their name, with older elements using the Latin version of their name. When two or more elements shared the same letter, the first two letters would be utilized. For example, gold (*Aurum* in Latin) was identified by the symbol *Au* while chromium and calcium were given the symbols *Cr* and *Ca*, respectively. Berzelius later expanded this system for the naming of compounds (see CHEMISTRY).

For some time chemists had observed that some groups of elements displayed similar characteristics. For example, both potassium and sodium reacted the in the same way with water. However, organization of these elements was hindered by the fact that little was understood about their chemical characteristics. Modern chemists recognize that elements have distinct atomic weights and valence structures (see ATOMIC WEIGHT, CHEMISTRY). While each of these were actively being studied by 19th century chemists, the data that they were deriving from their studies varied dramatically, thus complicating early efforts. The English chemist John Newlands made significant progress when he noticed that when the elements were arranged using both atomic weight and similar chemical characteristics, a pattern appeared that repeated every eight elements. This is sometimes called the law of octaves. Newlands's work was expanded upon in 1869 when the Russian chemist Dmitri Mendeleev arranged the elements by atomic weight. Mendeleev left blank spaces for undiscovered elements, which he named *eka* (Sanskrit for "one" or "first") elements since they were located on the table below known elements (see Figure 18). Thus there was an eka-aluminum, eka-silicon and so on for spots on the table where Mendeleev suspected that additional elements would be found later. This system of using the prefix *eka-* to indicate a spot on the periodic table that should contain an as yet unidentified element continues in modern chemistry. A similar table, but one that arranged the elements by chemical characteristics, was prepared by the German chemist Lothar Meyer (1830–1895) in 1870. Mendeleev believed that he could predict the chemical characteristics and atomic weights of these undiscovered elements based on the positions of the blanks on the table. This concept of looking for elements to fit the table resulted in the discovery of many elements (see Figure 18), such as gallium (1874), scandium (1879) and germanium (1886). This method of classifying elements serves as the basis for the modern version of the periodic table of the elements.

The next major advance in chemistry that aided the discovery of the elements began in 1814 with the work of the German physicist Joseph von Fraunhofer. Fraunhofer duplicated many of Isaac Newton's 17th century experiments with optics. He noticed that when a light was passed through a narrow opening and then a prism the resulting spectrum of colors contained

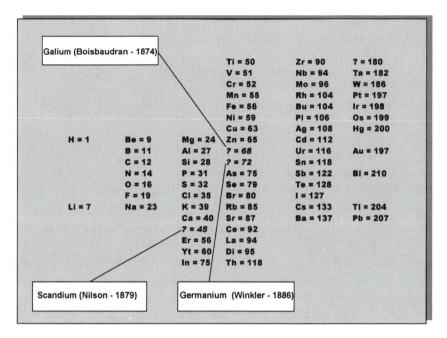

Figure 18. An illustration of Mendeleev's periodic table, indicating three of the eka-elements, and the investigators who eventually discovered them based on Mendeleev's predictions. Derived from Mendeleev's *Principles of Chemistry*, 1901.

a series of dark lines. What Fraunhofer had discovered are called spectral lines, sometimes called Fraunhofer lines (see ELECTROMAGNETIC SPECTRUM). Before the 19th century many investigators had observed that the **combustion** of compounds containing certain elements produced identifiable colored flames. By the mid-1800s many were using Fraunhofer's spectral lines to distinguish between different elements in a compound. In 1854 the American scientist David Alter (1807–1881) determined that each element had a unique set of spectral lines and proposed that this spectral fingerprint may be used to detect the presence of an element in meteors. The first new element that was detected using spectral analysis was cesium, by Gustav Robert Kirchhoff (1824–1887) and Robert Bunsen (1811–1899). Kirchhoff and Bunsen invented a spectroscope, an instrument used to generate and analyze spectral lines, specifically for the study of elements. In 1860, while observing the spectral lines of several known compounds, they discovered a substance that produced a spectrum consisting of twin blue lines. The named this new element cesium, from the Latin word for blue.

Kirchhoff followed with the discovery of rubidium in 1861. The development of spectral analysis played an important role in the discovery of a number of elements in the mid-19th century, including the discovery of the inert gases helium and argon.

At the end of the 19th century the discovery of radiochemistry played an important role in the study of the radioactive elements. Radioactivity is the result of an instability in the nucleus of an atom. This instability causes the nucleus to decay, and release particles, such as alpha and beta particles, or radiation, such as gamma radiation or x-rays. The majority of studies into electromagnetic radiation would be performed in the 20th century, but by the end of the 19th century scientists were recognizing that some elements had radioactive properties. The French physicist Antoine-Henri Becquerel had discovered in 1896 that uranium emitted gamma rays (see ELECTRO-MAGETIC SPECTRUM), but the prime investigator of radioactivity at the end of the 19th century was the Polish chemist Marie Curie (1867–1934), assisted by her French husband Pierre Curie (1859–1906). Marie Curie measured and described the radioactive properties of both uranium and thorium using a process called **piezoelectricity** (see ELECTRICITY). In her measurements of uranium's radioactivity she observed that the amount of radiation being produced was not consistent with the quantity of uranium in the sample. This led to the discovery of polonium in 1898, the first element to be discovered based on its radioactive properties. Marie Curie's work resulted in her being awarded two Nobel Prizes in chemistry (1903 and 1911), one of only two chemists to have received this honor.

By the end of the 19th century chemists had discovered or isolated 52 new elements, more than at any other time in the history of chemistry (see Table 2). Yet, of greater importance to the scientific community were the procedures that enabled the discovery of these elements. The procedures of spectral analysis and piezoelectricity play an important role in modern chemical investigations to this day. Furthermore, the discovery of radioactive elements, and the classification of their radioactive properties, resulted in extensive investigations into the structure of the atom in the 20th century, eventually resulting in the dawning of our current Nuclear Age. It can be said that the study of the elements in the 19th century marked the modernization of the science of chemistry into an important field of modern science.

Selected Bibliography

Asimov, Isaac. *Asimov's New Guide to Science*. New York: Basic Books, 1984.

Cobb, Cathy, and Harold Goldwhite. *Creations of Fire: Chemistry's Lively History from Alchemy to the Atomic Age*. New York: Plenum Press, 1995.

Krebs, Robert E. *The History and Use of Our Earth's Elements: A Reference Guide.* Westport, CT: Greenwood Press, 1998.

Weeks, Mary E., and Henry M. Leicester. *Discovery of the Elements,* 7th edition. Easton, PA: *Journal of Chemical Education,* 1968.

Elevator (1852–1880): Records show that elevators were used as far back as the 3rd century B.C.E. The earliest device was little more than a platform suspended by a rope over a pulley. The power to move these platforms was provided by either animals or humans and sometimes by water. For centuries these inventions had the ability to lift loads a relatively small distance but sufficed because of the relatively small height of pre–Industrial Revolution buildings. However, with the onset of the Industrial Revolution and the invention of new, improved building materials such as steel (see STEEL) it was becoming possible to build structures of greater height. Architects and construction engineers recognized that to move materials and people greater distances some sort of lifting device was needed. Improvement in the design and power output of the steam engine resulted in the construction of a number of steam-powered elevators in the early 19th century. Most these machines were used for industrial applications. However, although these machines were able to lift objects greater distances, the safety of the lifts was questioned for the movement of people. In most cases the platform was suspended by a single rope or cable, and if a mishap were to occur it lacked any safety protection. Some early elevators also used a hydraulic system to push the platform into the air. These were considerably more powerful machines, but they were limited in the height that they could move the platform and their relatively slow speed made them unattractive for moving people.

This situation changed in 1852 when the American inventor Elisha Graves Otis (1811–1861) invented the first safety elevator. Like many machines of the mid-19th century, Otis's invention was steam powered, which was not something new, but what he did incorporate into his design was a braking mechanism. The brake consisted of a spring and ratchet assembly that engaged only once the tension on the support rope of cable was released. Investors were a little skeptical of the device at first. However, in 1853 Otis made a convincing demonstration of his improved elevator at the New York Crystal Palace Exposition when he hoisted himself into the air and then instructed that the rope be cut. Sure enough, the platform descended only a few inches before the braking mechanism engaged. The first passenger elevator was installed several years later (1857) in a New York City department store. It could lift people and light cargo five floors in around a minute.

Several other improvements were made in elevator design in the 19th century. In 1880 the German inventor Werner von Siemens (1823–1883) invented the first electric elevators. Electricity was an important development of the mid-19th century (see ELECTRICITY, ELECTRIC POWER), and it proved a useful source of energy for the elevator. Others worked to improve the hydraulic systems for heavy-lift elevators. Although the elevator may seem to be a relatively minor invention of the 19th century, in reality when coupled to the invention of steel (see STEEL) it enabled the construction of taller buildings, also called skyscrapers. With the increase in office space that taller buildings provided, plus the convenience of being in close proximity to related businesses and services, the Western cities of the late 19th century witnessed a tremendous growth. Thus is may be said that, at least indirectly, the invention of the elevator enabled the formation of the modern urban environment.

Selected Bibliography

Karwatka, Dennis. *Technology's Past: America's Industrial Revolution and the People who Delivered the Goods.* Ann Arbor, MI: Prakken Publications, 1996.

Van Dulken, Stephen. *Inventing the 19th Century: 100 Inventions That Shaped the Victorian Age from Aspirin to the Zeppelin.* New York: New York University Press, 2001.

Embryology (1815–ca. 1881): Since the invention of the microscope in the 17th century scientists had demonstrated an interest in studying the earliest stages of life. One of the first scientists to examine the stages of embryonic development using a microscope was Marcello Malpighi. However, despite the discovery of the human sex cells during this time, there were a number of different ideas as to how an embryo was formed. Some, such as Malpighi, believed in *preformation*, meaning that miniature, complete individuals were preassembled inside each of the sex cells. Others of the time, such as the English scientist William Harvey (1578–1657), believed that the structures of the embryo gradually formed from simpler substances, a theory called *epigenesis*. In the 18th century the concept of pangenesis gained favor. *Pangenesis* was the idea that each individual part of the body contributed microscopic versions of itself to the forming embryo. Thus the hands of the body would contribute miniature hands, and so on. Around the same time the German physiologist Caspar Wolff (1738–1794) expanded on the idea of epigenesis when he suggested that all tissues form from unspecialized material. During the 19th century the study of embryology followed two main paths. First there was a group of scientists that were examining the process by which a fertilized egg becomes a functional organism. A second group was investigating the possibility that life could arise from inanimate matter. As with many areas of the biological sciences during this time, a series of significant advances set the stage for the development of modern theories in the 20th century.

The scientific community had debated for centuries the idea that life possessed the ability to spontaneously generate from nonliving material. In the 17th century the Italian scientist Francesco Redi (1626–1678) had demonstrated that maggots do not spontaneously form from rotting meat. Additional experiments in the 18th century by John Needham (1713–1781) and Lazzaro Spallanzani (1729–1799) provided even more evidence against spontaneous generation. Yet until the 19th century many scientists still held that some form of spontaneous generation was possible among microscopic organisms. Even the founders of the cell theory believed that it was possible for new cells to form by crystallizing from the inanimate interior contents of the cell (see CELL THEORY). In 1815 the German chemist Franz Schultze designed an experiment in which a nutrient broth was first boiled and then placed into flasks. One of these flasks was left open to the air, while the other was isolated from the external environment by a solution of sulfuric acid. As predicted, the flask left open developed a healthy growth of microbes, and the isolated flask did not. Perhaps the most conclusive evidence against spontaneous generation was provided by Louis Pasteur, the developer of the germ theory of disease (see GERM THEORY). In his experiments on fermentation (see FERMENTATION) Pasteur demonstrated the many microbes are present in the atmosphere and thus do not spontaneously generate in a sealed flask (see GERM THEORY for diagram). The final evidence against spontaneous generation was provided by the Irish physicist John Tyndall (1820–1893). Tyndall used the same basic experimental procedure as Pasteur, in which a series of nutrient-rich flasks were exposed to the environment while others remained sealed. But what Tyndall noted was that the level of "spontaneous generation" that was observed in the open flasks was dependent on the air quality. Sterile rooms differed in the level of contamination when compared to open laboratories. When combined with the germ theory of disease it was evident that airborne microbes were responsible for spontaneous generation.

With the popularity of spontaneous generation greatly decreasing throughout the 19th century, a number of biologists began to focus on detailing the stages of embryonic development. One of the most important of these was the Russian biologist Karl Ernst von Baer (1792–1876). Baer is credited with discovering the human ovum, or female sex cell. For some time scientists had been examining tissue from the human reproductive system in an attempt to localize the ovum, but it was not until Baer's research (ca. 1828) that the elusive cell was localized from tissues of the ovary. Baer made a far more significant contribution to the science of embryology: the development of the germ layer theory. In the 18th century the German embryologist Caspar Wolff had described a series of tissue layers in the development of plant embryos. Baer examined these layers in animal embryos and concluded that distinct layers formed early in development that subsequently gave rise to the major structures of the body. He described four of these germ layers and made one of the first suggestions of what is called the biogenic law. This states that all higher animals pass through embryonic developmental stages that resemble those of lower animals. This idea became an important premise of the field of evolutionary and comparative embryology (see later). Furthermore, von Baer recognized that during development of an embryo the more general structures formed first, followed by the more specialized structures. He is also credited with the discovery of the notochord (see also CELL THEORY), an important structure in evolutionary history of **vertebrate** organisms.

The germ layer theory was later refined by two important individuals. The first of these was the German scientist Robert Remak, who also made important contributions to the study of cell division (see CELL DIVISION). Remak recognized that Baer's four germ layers were actually only three and named them the endoderm (inner layer), mesoderm (middle layer) and ectoderm (outer layer). His studies (ca. 1845) further defined the role of each of these layers in the developed organism. Remak suggested that the ectoderm was responsible for the formation of the skin and nervous systems; the mesoderm formed the muscular, excretory and skeletal systems; and the endoderm was responsible for the development of the digestive tract and notochord. It is now recognized that there are some minor deviations in this pattern, but for the most part Remak's ideas form the basis of modern embryological thinking.

The study of embryonic development also influenced the 19th century investigators of evolution, most notably the English naturalist Charles Darwin (1809–1882). In Darwin's 1859 publication of *The Origin of Species* he refers indirectly to the biogenic law (see THEORY OF NATURAL SELECTION), and it is known that Darwin was influenced by the work of Baer and Remak. Around this same time a new field of embryology arose that examined the evolutionary relationships of the embryonic development of many animal **species**. Some of the most groundbreaking work in this field was conducted by Alexander Kowalewsky (1840–1901), who demonstrated that **invertebrate** animals follow the same basic embryonic development as vertebrate animals. In his studies of the invertebrate **Chordate** species *Amphioxus*, he established that the embryonic development of this species indicates that the formation of the notochord in this species is similar to that of all other chordate animals and thus it possibly represents a direct descendent of the first organisms with the notochord. Johannes Müller (1801–1858) and Heinrich Rathke (1793–1860) recognized that the gill slits and gill arches of fish are present in all embryonic forms of higher vertebrates, specifically the mammals and birds. Furthermore, some structures of the higher vertebrates, namely the bones of the jaw, are formed from the gill arches during embryonic development. Still others, such as T. H. Huxley (1825–1895) and Ernst Haeckel recognized that animals could be classified by minor differences in embryonic development. Haeckel proposed that "ontogeny recapitulates phylogeny," meaning that the embryonic development of an organism (ontogeny) represents a recorded biological history of the evolution of the species (phylogeny). In Haeckel's examination of sponges he invented some of the modern terms, such as blastula and gastrula, used to describe the stages of embryonic development (see Figure 19).

By the end of the 19th century a tremendous amount of research was being conducted in the field of evolutionary embryology. In the process a much clearer picture was being established of the evolutionary history of the animal kingdom. Published works by Louis Agassiz (1807–1873) and Francis Balfour (1851–1882) summarized the knowledge that was being accumulated in this field and helped established the role of embryology in evolutionary studies. The science of biology entered the 19th century debating the basic formation of the embryo and the role of spontaneous generation but exited the century with the foundations of some of the most important biological concepts of the times. The study of embryology in the 20th century would build significantly on the work of these pioneers. New molecular techniques in DNA

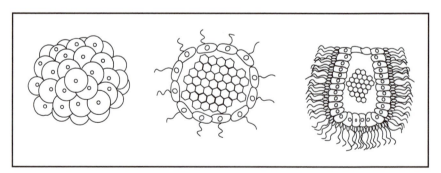

Figure 19. An illustration of Haeckel's work on embryology. The left stage is a group of undifferentiated cells. The middle stage depicts the blastula stage of embryonic development, and the right is the gastrula stage.

analysis and computer modeling would add to the understanding of evolutionary embryology. Still other scientists now examine the positive role of nutrition on embryonic development and the negative influence of man-made chemicals, specifically those associated with birth defects. It is now possible to screen a developing embryo for genetic problems, and possibly in the future we may be able to correct them through gene therapy. Recently there has been a renewed interest in embryology from the medical community with regard to stem cells, which are undifferentiated cells that mature to form a wide variety of different cell types. Some suggest that embryonic stem cells may be able to reform almost any type of tissue, including nerve cells, and thus hold promise as a possible solution for transplants and disease therapy. Although the study of embryology in the 20th and 21st centuries has progressed tremendously, the terminology and principles that it employs strongly indicate the importance of these preliminary 19th century studies.

Selected Bibliography

Mayr, Ernst. *The Growth of Biological Thought: Diversity, Evolution and Inheritance.* Cambridge, MA: Harvard University Press, 1982.

Needham, Joseph. *A History of Embryology.* New York: Abelard-Shuman, 1959.

Ronan, Colin A. *Science: Its History and Development Among the World's Cultures.* New York: Facts on File Publications, 1982.

Serafini, Anthony. *The Epic History of Biology.* New York: Plenum Press, 1993.

Singer, Charles. *A History of Biology to about the Year 1900: A General Introduction to the Study of Living Things.* Ames IA: Iowa State University Press, 1989.

Enzymes (1811–1897): Enzymes function as the organic catalysts of chemical reactions. This means that they accelerate the rate of a reaction, usually by reducing the amount of energy required for the product to be formed. Because many chemical reactions either will not occur spontaneously or proceed at a slow rate, enzymes are crucial for the metabolic properties of biological organisms. However, enzymes are not confined to biological systems and may participate in almost any form of chemical reaction. The advances in chemistry during the 18th century had made it possible for scientists to quantify chemical reactions, a process now called enzyme kinetics. In the 19th century the study of enzymes, also called soluble ferments, and the process of fermentation were intertwined, with each attempting to explain how biological organisms convert raw materials into the organic materials needed for the life of the organism.

The German chemist Gustav Kirchhoff, recognized for his innovative methods of discovering new elements (see ELEMENTS), performed one of the first studies (ca. 1811) of accelerated chemical reactions in the 19th century. Kirchhoff's experiment examined the breakdown of starch into simple sugars. In his studies Kirchhoff noticed that when sulfuric acid was added to the reaction, the rate of starch breakdown was significantly enhanced. Furthermore, the sulfuric acid itself was unchanged during the process of the reaction. This knowledge initiated a series of studies of similar materials that increase the speed of the reaction but remain chemically unaltered throughout the process. In 1833, two chemists, Anselme Payen (1795–1871) and Jean Persoz (1805–1868) conducted an experiment in which they rinsed germinating barley seeds with alcohol. When this extract was placed back into a water environment, it possessed the ability to break down starch into simple sugars. Payen and Persoz named this compound *diastase*. Although they did not know it, Payen and Persoz had isolated the first enzyme. Similar experiments were conducted by Theodor Schwann in 1836. Schwann's experiments involved isolating the solutions produced by the lining of the stomach by filtering cells of the lining through a series of paper and cloth barriers. The resulting solution had the ability to dissolve proteins in a test tube. Schwann named this compound *pepsin* and observed that it had the ability to break down materials from a wide variety of sources, including egg whites, muscles and blood. The term *soluble ferments* was frequently applied to pepsin and diastase because they did not require a living cell to conduct the reaction. But these types of reactions were not limited to organic material. For example, the metal platinum had the ability to easily combine free hydrogen and oxygen to form water, a reaction that normally only occurs only at high temperature. In 1837 the renowned Swedish chemist Jöns Berzelius, known for his work in discovering elements (see ELEMENTS) applied the word *catalyst* to these compounds and defined them as substances that promote the rate of a chemical reaction without being used themselves in the reaction. He also suggested that life was the result of thousands of these reactions, a foreshadowing of the findings of 20th century biochemists.

At about the same time as these studies were being done, a second group of scientists was investigating the biological process of alcohol fermentation. In the process of fermentation scientists recognized that the organic material found in grapes transformed into alcohol. Investigators of fermentation, such as Justis von Liebig and Louis Pasteur, believed that all catalytic reactions involving organic molecules were the property of the living cells themselves, and not a separate chemical. However, they differed in their explanations of the cellular process that causes fermentation (see FERMENTATION). The work of Pasteur and Liebig appeared for a while to be at odds with the studies of soluble ferments conducted by Schwann, Peyen and Persoz. However, in 1860 Moritz Traube (1826–1894), a Polish chemist, attempted to synthesize these ideas into a single workable theory. Traube suggested that the reactions of fermentations and the soluble ferments were in fact similar in many ways and differed only in the location in which they occurred (inside or outside the cell). He then went a step further and proposed that the reactions of these catalysts were not unique to lower organisms, but also could be used to explain the chemical processing of material in higher-level organisms. A partial proof of Traube's ideas occurred a few years later when the French chemist Marcelin Berthelot (1827–1907) isolated a compound from yeast that had the ability to break down the sugar sucrose into its two structural components: glucose and fructose. This enzyme was named invertase and was important because it was the first catalyst known to have been isolated from within a living cell that also demonstrated activity outside the cell. The action of invertase was similar in nature to the soluble ferments and thus Berthelot suggested that living cells may be the source of all biological catalysts.

It would appear that Berthelot's experiment should satisfy both major camps of thought, but no one to this date had demonstrated the action of a catalyst **in vitro**, or outside a living cell. Therefore it remained a possibility that the cell produced soluble ferments to facilitate reactions outside the cell but that a different mechanism was responsible within a cell. In 1897 Hans (1850–1902) and Eduard Buchner (1860–1917) demonstrated that the process of fermentation could be performed in a test tube, without living cells being present. To do this the Buchner brothers isolated an enzyme they called zymase from yeast cells. When placed in a test tube containing a high concentration of sugar, the zymase solution conducted the same fermentation reactions that had been observed to occur within the living cells, meaning that alcohol (ethanol) and carbon dioxide were being produced. Thus, they concluded that the zymase enzyme must be responsible for the same reaction both inside and outside the cell. Although this idea was not widely accepted at the time, over the next several years additional experimental evidence substantiated their findings.

The findings of these 19th century scientists set the stage for the 20th century study of biochemistry, or the chemistry of living organisms. In the early 20th century biochemists began to realize that the overwhelming majority of chemical reactions within the cell are associated with one or more enzymes. Furthermore, these enzymes may be regulated by coenzymes, or metabolic assistants. In 1913 an attempt was made to establish a mathematic relationship between the amount of substrate (starting material) present in a reaction and the overall rate of the reaction. This formula is called the Michaelis–Menten equation after the two German chemists, Leonor Michaelis (1875–1949) and Maud Menten (1879–1960), who developed it. The Michaelis–Menten equation is still used today in analyzing the rate of enzymatic reactions. Many other studies of enzymes were undertaken in the 20th century, one of the greatest of which involved the process of enzyme crystallization. In 1926 the American scientist James Sumner (1887–1955) was able to crystallize the urease enzyme for the first time. Crystallization allows a biochemist to study the structure of an enzyme in greater detail. From this work Sumner discovered that enzymes were proteins, as many had previously thought but never proved.

The term *enzyme* was first used to describe the biological catalysts in 1878 by Wilhelm Kuhne (1837–1900). This term is now frequently used to describe all aspects of chemical catalysts. Enzymes are now generally classified according to what compounds, or substrates, they react with. For example, enzymes such as diastase and zymase that react with carbohydrates belong to a group of enzymes called the amylases, from the Greek word for sugar. Those enzymes that react with proteins are commonly called proteases. By convention, the *-ase* suffix is commonly used to designate the enzyme for a particular molecule.

For most of the 20th century biologists considered that all enzymes were proteins, primarily because the three-dimensional shape of proteins makes them ideal molecules to catalyze chemical reactions (see BIOMOLECULES). However, by the end of the 20th century evidence was accumulating that some forms of ribonucleic acid, or RNA, may also have enzymatic functions. Understanding how these enzymes work may shed light on the formation of the first cells in the Earth's oceans. However, the majority of enzymes are proteins. Biochemists now devote a significant amount of time not only to the study of enzyme kinetics but also the genetic mechanisms of enzyme regulation and the relationship between enzyme function and health. The advances in modern biochemistry are a result of the foundation of understanding developed in the 19th century.

Selected Bibliography

Dressler, David. *Discovering Enzymes*. New York: Scientific American Library, 1991.
Fruton, Joseph H. *Molecules and Life: Historical Essays on the Interplay of Chemistry and Biology*. New York: John Wiley & Sons, 1972.
Serafini, Anthony. *The Epic History of Biology*. New York: Plenum Press, 1993.
Singer, Charles. *A History of Biology to about the Year 1900: A General Introduction to the Study of Living Things*. Ames IA: Iowa State University Press, 1989.

F

Fermentation (ca. 1800–1897): Although there are many different chemical forms of chemical fermentation reactions, the most common of them is the **anaerobic** breakdown of sugars to form alcohol. The fermentation of grapes and malts to form alcoholic beverages is recorded by a number of early civilizations, which used the product for a variety of medicinal, recreational and religious purposes. The scientific basis of fermentation was studied by both the early Chinese and ancient Greek cultures, although these early groups lacked the ability and methods to effectively understand the biological relationship of the process. The invention of **distillation** around the 12th century allowed for the isolation of the alcohol from the reaction. The chemical study of fermentation began in the 17th century with the start of the Scientific Revolution. Around this time many early chemists, called alchemists (see CHEMISTRY), attempted to explain the process as a simple **transmutation** of matter. A century later the influential French chemist Antoine Lavoisier devised an experimental system to monitor the movement of carbon atoms during the fermentation reaction (see later). He showed that there was a conservation of mass within the reaction, that is, the carbon in the original glucose could be accounted for in the end products of ethanol and carbon dioxide. In the very early years of the 19th century the French chemist Joseph Gay-Lussac developed the following chemical formula for the reaction:

$$C_6H_{12}O_6 \rightarrow 2CO_2 + 2C_2H_5OH$$

Glucose $\rightarrow$ carbon dioxide and ethanol

As the century began, however, it was unclear as to the mechanism by which the reaction occurred, specifically the role of yeast in this process. The study of fermentation in the 19th century was directly connected with several other evolving areas of the biological sciences, including the study of spontaneous generation (see EMBRYOLOGY), the study of enzymes (see ENZYMES) and the development of the germ theory of disease (see GERM THEORY). Researchers in each of these areas developed independent theories about the nature of fermentation, each one of which made a contribution toward understanding the scientific basis of fermentation.

The chemical reaction shown above will not occur without the presence of yeast. Yet during the reaction the amount of yeast does not decrease, thus it is not being used up during the reaction. At the start of the 19th century, scientists considered fermentation to be a purely chemical process that was independent of the biological system. In the early decades of the 19th century there was considerable interest in describing the action of catalysts, also called soluble ferments (see ENZYMES). Soluble ferments such as the digestive enzyme pepsin and the carbohydrate enzyme diastase had been isolated outside the cell and did not appear to require the presence of a living cell to function. To many scientists, the yeast in the reaction simply provided the catalysts necessary for the production of alcohol. By the 1830s several theories were being developed in an attempt to explain the role of the yeast in this biochemical process. For example, around 1839 the German chemist Justus von Liebig attempted to explain the process of fermentation as a mechanical-chemical reaction. Liebig suggested that the motion of the atoms within the yeast cell effectively made it unstable. The close proximity of the yeast to the sugar caused the vibrations of the yeast to be passed on to the atoms of the sugar. This in turn resulted in a change in the molecular configuration of the sugar, which formed ethanol (alcohol) and carbon dioxide. Furthermore, during the process the yeast literally shook itself apart and degraded. Unfortunately this degradation of the yeast goes against the rules for a catalyst (see ENZYMES), but Liebig's idea sounded plausible to many.

In his explanation Liebig did not initially accept that yeast was a living organism. However, within just a few years of Liebig's work Theodor Schwann, codeveloper of the cell theory (see CELL THEORY), performed studies of the reaction, using a microscope, that suggested not only that the yeast was alive but also that a definite relationship existed between the growth of the yeast and the formation of the alcohol. It was becoming obvious that fermentation was not a simple chemical reaction, but one that involved the influence of a biological system. Although yeast could be observed growing during the reaction, it was still believed by many that fermentation was a purely chemical process.

This belief changed in 1856 when the influential French chemist Louis Pasteur became interested in fermentation. Pasteur was invited by a local winery to investigate a problem with their fermentation process (see Figure 20). Instead of producing ethanol, certain stocks of wine were producing large

Figure 20. An illustration of Pasteur at work in his laboratory. Besides his development of the process of pasteurization (see GERM THEORY), Pasteur was frequently called on to solve other industrial problems during the 19th century. (Hulton Archive/Getty Images)

quantities of lactic acid, resulting in a very sour product. When Pasteur examined cultures of the samples that produced lactic acid under a microscope, he noticed that they contained a different microbe from those that were producing ethanol. From this observation he concluded that the products of the reactions, lactic acid and ethanol, were the result of the metabolic activity of two different organisms. This was an important finding in itself, but Pasteur did not stop with this one observation. Instead, he developed an experimental system in which he took a small sample from both the lactic acid and the ethanol stocks and placed them individually in solutions that contained sugar and other materials, such as minerals, which were needed for life (see GERM THEORY for figure). To ensure that the solution did not contain microbes before the reaction took place, he heated it. This heating procedure is called pasteurization and is used throughout modern industry to destroy microbial contaminants. Pasteur's experimental system allowed him to grow relatively pure cultures of each microbe for study. From this experiment he made several important conclusions. First he proposed that the yeast cells metabolized the sugars and other materials in the flask to produce specific products, namely ethanol and lactic acid. Second, he suggested that the nature of the medium in which the yeast was grown influenced the formation of the product. Of a far greater importance, though, was his discovery that the initial source of the bacteria was the air itself. This discovery led Pasteur into the study of the germ theory of disease, for which he is widely recognized (see GERM THEORY).

Although Pasteur was correct in his conclusions that the process of fermentation was the result of the biological activity of the yeast, he believed that the reaction must be conducted by an intact cell. In other words, he thought that fermentation was an inherent property of the living organism and could not be separated from the cell and carried out in a test tube. However, a series of events in the study of enzymes demonstrated that this view was not entirely correct. The discovery of the soluble ferments (enzymes) diastase and pepsin outside the cell demonstrated that certain biochemical reactions did not require the presence of a living cell (see ENZYMES). The problem with both diastase and pepsin was that they were secretions of the cell, designed to function in the extracellular environment. However, in 1860 Marcelin Berthelot isolated an enzyme called invertase from a living cell. This enzyme maintained its chemical activity after being removed from the cell (see ENZYMES). Thus it was becoming plausible that fermentation was actually a chemical reaction being conducted within the cell, a synthesis of Liebig's and Pasteur's ideas. The proof that enzymes are involved in fermentation was provided in 1897 by Hans Buchner, who discovered the enzyme responsible for the process of fermentation, which he called zymase.

The study of fermentation in the 19th century had progressed from the initial view of its being purely chemical reaction, to the understanding that it was a process of a biological system and then finally to the realization that it was a biochemical reaction. The study of fermentation, along with the study of enzymes, was the start of the science of biochemistry, or the chemical reactions of living cells. At the end of the 19th century, biologists were recognizing that living organisms transform the nutrients in their environment into living material via chemical pathways. In addition, these investigations into the process of fermentation provided important insight into the role of bacteria, which led to the development of the germ theory of disease and modern medical practices.

Selected Bibliography

Dressler, David. *Discovering Enzymes*. New York: Scientific American Library, 1991.
Fruton, Joseph H. *Molecules and Life: Historical Essays on the Interplay of Chemistry and Biology*. New York: John Wiley & Sons, 1972.

Geison, Gerald L. *The Private Science of Louis Pasteur*. Princeton, NJ: Princeton University Press, 1995.

Serafini, Anthony. *The Epic History of Biology*. New York: Plenum Press, 1993.

Singer, Charles. *A History of Biology to about the Year 1900: A General Introduction to the Study of Living Things*. Ames IA: Iowa State University Press, 1989.

Fossil Fuels (1800–1859): The term *fossil fuel* is used to describe compounds that are made of long chains of carbon and hydrogen molecules called hydrocarbons (see CHEMISTRY, ORGANIC CHEMISTRY). They are the remains of organic creatures, most of which lived during the Carboniferous period of the Earth's history (see PALEONTOLOGY for diagram). The **potential energy** stored within these bonds can be easily released by the process of combustion. Fossil fuels were not actually a discovery of the 19th century, but the development of methods to effectively extract and modify the fossil fuels into substances that could power the machines of the Industrial Revolution was to the result of the actions of a group of 19th century inventors. Since the early times of civilization the power to do work was provided by the brute force of animal and human labor. In many cultures the power of moving water and wind was also harvested. Things changed significantly in the 17th century with the invention of the first modern industrial steam engine by the English inventor Edward Somerset (1601–1667). Power for the steam engine was typically supplied by wood, and later the fossil fuel coal. Improvements in the design of the steam engine in the 18th and 19th centuries increased its popularity, and by the mid-19th century steam power was dominating the industrial world (see STEAM ENGINES). Fossil fuels also played an important role in the early forms of artificial light and the development of the internal combustion engine during the 19th century. By the end of that century the fossil fuels had proved versatile enough to satisfy the majority of the needs of the Industrial Revolution.

Coal represents one of the earliest forms of fossil fuels to be used by humans. In some areas of the world, coal deposits are found naturally on the surface, and early Chinese and Native American cultures frequently used coal for a variety of purposes. By the 19th century, coal and a similar compound called peat were being mined extensively in both Europe and North America, where they provided a plentiful, but highly polluting, source of energy. The first refinement of fossil fuels for specific purposes was most likely coal gas. At the end of the 18th century the English inventor William Murdock (1754–1839) discovered that during the processing of coal to form coke (see STEEL), the gas that was produced was also inflammable. In 1800 Murdock was successful in capturing the coal gas and using it for artificial lighting.

Although other fuels were used during the 19th century for artificial lighting (such as kerosene and whale oil) the use of coal gas in major cities did much to transform the nature of urban living and remained the primary form of artificial lighting until the invention of electric light and electric power later in the century (see ELECTRIC LIGHT, ELECTRIC POWER).

Another refined fuel of the 19th century was kerosene. Kerosene was first produced in 1853 by the English doctor Abraham Gesner (1797–1864). Natural fossil fuels may occur in a wide variety of forms, including gases, liquids and solids. Some solid forms are hard, like coal, but others may have a waxlike consistency, such as shale. Gesner developed a way to use heat to separate the liquid and solid forms to make kerosene, a highly flammable liquid. Kerosene was a more efficient fuel than whale oil and produced less smoke, so it immediately gained popularity for use in lamps.

The term *petroleum* refers to liquid hydrocarbons, or what is commonly called oil. Various forms of oil, from both whales and plants, have historically been used as fuel. In most cases petroleum oil is found deep underground, where it is trapped between layers of dense rock. However, owing to faults and variations in the rock layers, petroleum sometimes seeps to the surface where it can easily be collected. In some cases shale deposits may also hold oil. Petroleum oil has been used for centuries to waterproof and protect boat hulls. In some cases it was used as a weapon of war during both naval and land combat. In certain areas the petroleum that seeps to the surface separates into lighter and heavier components. The lighter component, called naphtha, was frequently used in lamps in the same manner as kerosene. However, in most areas of the world petroleum deposits were inaccessible. Because petroleum and kerosene are both hydrocarbons, chemists recognized that it would be possible to manufacture large of amounts of the popular fuel kerosene if there were a consistent source of oil.

A solution to the problem was provided in 1859 by the American Edwin Drake (1819–1880). Drake is recognized as the first person to drill for oil, although he did not either invent the drilling process or create the concept that there may be oil deposits located underground. For some time settlers of the American west had used simple drilling rigs and wells to obtain water and salt brine. The idea to drill for underground oil came from the president of a Pennsylvania oil company that was actively removing ground-level oil from a pool outside Titusville, Pennsylvania. After a series of attempts, Drake constructed a 50-foot wooden structure that housed a steam-powered drilling apparatus. After drilling down 69 feet, the device struck petroleum oil and in the process created the first oil well. Production from this well was initially very slow, which actually benefited the company since the first large-scale refinery would not be built for almost two years. Still, the well had proved that oil could readily be obtained, and in the very near future the market would increase.

The invention of the electric light severely reduced the need for kerosene and other petroleum products. By the mid-1840s liquid hydrocarbons were becoming a popular fuel among the inventors of the internal combustion engine. Its ability to lubricate moving parts and retain its lubrication properties under heat made it popular among the early inventors of the internal combustion engine (see INTERNAL COMBUSTION ENGINE). Refining petroleum yields kerosene and another, lighter component called gasoline. Although many of the early engines used other hydrocarbons or alcohols as fuel, gasoline was easily mixed with air and ignited in the cylinder. By the end of the century the internal combustion engine had been incorporated into the design of the horseless carriage, producing the first automobiles (see INTERNAL COMBUSTION ENGINE). In most cases gasoline was the fuel of choice for these engines.

The primary problem with gasoline was that it was difficult to control the combustion of the gas as the pressure increased inside the engine's cylinder. Initially this was not much of a problem, because automobiles were primarily a curiosity, but as the vehicles gained popularity the demand for power and an aesthetic need to eliminate the knocking sound caused by premature ignition in the cylinder resulted in the addition of antiknock compounds. One of these, introduced in the 1920s, was lead. Unfortunately lead is a toxic compound, and its use as a fuel additive has been eliminated or limited in most areas of the world. Other compounds, such as methyl tertiary butyl ether (MTBE), have recently been invented, but the question over the carcinogenic nature of these compounds raises questions about their long-term use.

The rise of the petroleum industry in the 19th century and the widespread use of all fossil fuels in the 20th century have created a significant pollution-related problems. When gasoline is burned in an engine the waste materials are expelled into the atmosphere. In some cases impurities, such as sulfur, exist in the gasoline. These mix with water in the atmosphere and return to Earth as pollution such as **acid rain**. Another problem is that the combustion of fossil fuels produces a significant amount of carbon dioxide. Although carbon dioxide is needed for plant growth, it was recognized in 1863 by the Irish scientist John Tyndall to be responsible for trapping the heat of the earth, a process that results in what is called the greenhouse effect. In recent years it has been estimated that 7 billion tons of carbon dioxide gas are released into the atmosphere annually by the burning of fossil fuels. This excess is believed to have begun a process of global warming as a result of the greenhouse effect. For this reason, plus the fact that there is only a finite amount of fossil fuels available, many scientists are suggesting a return to alternative energy sources. Unfortunately, in this case the 19th century scientists did their job too well, and the industrialized world, at least for the moment, is tied to fossil fuels for the production of most of the products of modern civilization.

Selected Bibliography

Asimov, Isaac. *Asimov's New Guide to Science*. New York: Basic Books, 1984.

Forbes, R. J. *Studies in Early Petroleum History*. Leiden, Netherlands: E.J. Brill, 1958.

Karwatka, Dennis. *Technology's Past, Vol. 2: More Heroes of Invention and Innovation*. Ann Arbor, MI: Prakken Publications, 1999.

Owen, Edgar Wesley. *Trek of the Oil Finders: A History of Exploration for Petroleum*. Tulsa, OK: The American Association of Petroleum Geologists, 1975.

G

Gas Laws (1808–1865): For chemists and physicists, a gas may be defined as the physical state of matter in which interaction between the particles is minimal and the resulting form lacks a defined shape. The study of the chemical properties of gases represents one of the areas of greatest scientific concentration in the first half of the 19th century. The study of gases was not unique to the 19th century. In the 17th century the English chemist Robert Boyle had established the fundamental relationship of pressure to gas volume, now called Boyle's law. In the 18th century the French physicist Jacques-Alexandre Charles (1746–1823) determined that the volume of a gas is dependent on the absolute temperature of the gas. This is called Charles's law, although it would undergo some revision in the 19th century before being stated in its final form. However, it was not until the beginning of the 19th century and the subsequent interest in atomic theory that many of the fundamental laws of gases were established. In the early years of the 19th century scientists were redefining the basic principles of atomic theory, much of which had been in existence since the time of the ancient Greeks. By 1810 John Dalton had established the fundamental principles of atomic theory, which stated that all matter is made of atoms, that all atoms of a given element have the same, unique mass, and that compounds consist of atoms in whole-number ratios (see ATOMIC THEORY). Much of the work in early atomic theory, including that of Dalton, was performed on gases such as carbon dioxide and nitrogen dioxide. The individual atoms of the gas were not bound to one another, as was the case with solids and liquids, and gases responded faster to temperature changes, a common variable in early atomic studies. Dalton's findings had a strong influence on several other scientists who were studying the properties of gases.

One of these was the French chemist Joseph-Louis Gay-Lussac. Gay-

Lussac made a number of important contributions to 19th century science (see also BIOMOLECULES). In the area of gas laws he was responsible for perfecting Charles's law of the 18th century. Whereas Charles had merely defined that a relationship existed between the temperature of a gas and its volume, it was Gay-Lussac who determined the precise mathematical relationship. Gay-Lussac determined that as a gas changed temperature by 1°C, its volume changed by a factor of 1/266.66. Another contribution of Gay-Lussac was the law of combining volumes. This was basically an extension of the work of Dalton's atomic theory to gases (see ATOMIC THEORY). It states that the when gases are involved in a chemical reaction, their volumes are always expressed in small, whole-number ratios. The work of Boyle, Charles and Gay-Lussac is frequently referred to as ideal gas laws because they define the mathematical relationship among temperature, volume and pressure. This can be expressed as:

$$PV = nRT$$

where P represents pressure, V equals volume and T is temperature. The n term refers to the number of moles of gas (see later) and R is a gas constant (8.314 joules).

Like Gay-Lussac and Dalton, the Italian chemist Amedeo Avogadro was very active in the early 19th century scientific community. His most important contributions were in the development of 19th century ideal gas laws. Avogadro was interested in defining the relationship between the number of atoms and the volume of a gas. This was not a unique area of study; Dalton had considered in his study of atomic theory (see ATOMIC THEORY) that gases with equal volumes would contain equal numbers of atoms. However, Dalton did not pursue this idea further. However, Gay-Lussac's law of combining volumes renewed interest in this idea. Avogadro noted that since a relationship existed between volume and the ratio of gases in a chemical reaction, there must be some correlation between the number of atoms and volume. Avogadro then stated that *at the same volume and pressure, equal volumes of gases contain the same number of molecules*. Furthermore, this relationship is independent of the physical and chemical properties of the gases. When developed this was initially called Avogadro's hypothesis (1811), but it is now widely recognized as Avogadro's law.

An important aspect of Avogadro's law is the use of the term *molecule* instead of atom. Dalton had believed that when atoms combined they did so in the lowest possible ratio of atoms (see ATOMIC THEORY). For example, water consisted of a single atom of oxygen and hydrogen. However, Avogadro noted that if his ideas on the relationship between the volume of a gas and the number of atoms were true, then the weights of the molecules of the gases must equal the ratios of the densities of the gases at the same temperature and pressure. In other words, if gas A has a density twice that of gas B, the atomic mass of the particles in gas A is also twice that of those comprising gas B. This is now known as Avogadro's hypothesis. In 1814 the French physicist André Ampère independently confirmed Avogadro's work. Avogadro used this reasoning to demonstrate that water consists of two parts hydrogen to one part oxygen, and proposed that this method may be used as an improved method of determining atomic weight (see ATOMIC WEIGHT). Furthermore, using this principle Avogadro distinguished between individual atoms and a *molecule* (*Latin*: "small mass"), a term that he used to name compounds comprising more than one atom. Avogadro noted that molecules of gases such as oxygen and nitrogen were diatomic, or consisted of two atoms. Unfortunately Avogadro's ideas were ahead of his time, and owing to a number of factors, including the political climate of the early 19th century, his ideas remained unrecognized by the chemical community for almost five decades.

Later in the 19th century several developments in the scientific world renewed interest in Avogadro's work. In 1860 the Italian chemist Stanislao Cannizzaro (1826–1910) discovered that Avogadro's hypothesis would allow him to accurately determine the molecular weight of gases and that this principle could be applied to the study of solids and their specific heat. This helped establish the practice of quantitative chemistry and found widespread application in the study of organic chemistry (see ORGANIC CHEMISTRY). In 1865 the Austrian chemist Johann Loschmidt (1821–1895) used Avogadro's hypothesis to calculate the number of atoms in a given volume of gas. It had previously been determined that hydrogen had a molecular weight of 1. Hydrogen gas is diatomic and, under ideal conditions, 2 grams of hydrogen gas occupy a volume of 22.4 liters. Using Avogadro's principles and the developing ideas on the kinetic theories of gases (see KINETIC THEORY OF GASES), Loschmidt was able to determine that 22.4 liters of hydrogen gas contained six hundred billion trillion (more exactly, 6.023×10^{23}) molecules. Because this information was derived from Avogadro's discoveries, it is called Avogadro's number or constant. Chemists now use the term *mole* to define the weight of an element or compound that contains exactly this number of atoms. For example, 2 grams of hydrogen gas equals one mole. In addition, 22.4 liters of hydrogen gas weighs 2 grams, or one mole. This constant is widely used in chemical investigations and represents an important advance in the study of quantitative chemistry.

The study of gases in the early part of the century had focused on their properties under the ideal conditions of constant pressure and temperature. However, ideal conditions are the exception in the natural world, and as the century progressed scientists began to study the behavior of gases under varying conditions. This led to the establishment of several principles on the kinetic theory of gases (see KINETIC THEORY OF GASES). These studies were a significant breakthrough for chemistry and physics because they represented a change from simply identifying the physical laws of nature to applying them to the study of gases under natural conditions. In doing so, the next generation of chemists would rely significantly on the scientific principles established by Dalton, Gay-Lussac and Avogadro in the early decades of the 19th century.

Selected Bibliography

Gillispie, Charles (ed.). *Dictionary of Scientific Biography*. New York: Charles Scribner's Sons, 1970.

Krebs, Robert E. *Scientific Laws, Principles and Theories: A Reference Guide*. Westport, CT: Greenwood Press, 2001.

Purrington, Robert D. *Physics in the Nineteenth Century*. New Brunswick, NJ: Rutgers University Press, 1997.

Geological Time (ca. 1830–1898): The scientific community has had a long-term struggle with the age of the Earth, primarily because of the absence of a mechanism for measuring the passage of time outside the boundaries of human record. Without a scientific mechanism for measuring the ages of rocks and the solar system, the age of the Earth had been established by examining the chronological records of civilization. In Western cultures these analyses primarily relied on examinations of time as recorded in the Bible. For example, in 1650 James Ussher, a Catholic Bishop, used the record of births and deaths within the Bible to calculate that the Earth was formed in 4004 B.C.E. Later religious historians used this method to calculate the exact hour and day of creation. The English astronomer Edmund Halley (1656–1742) attempted a more scientific approach during the late 17th century. As part of his research into the **hydrologic cycle** Halley reasoned that initially the oceans on the planet should have been freshwater. He presented the idea that the age of the Earth could be determined by first measuring the increase in the salinity of the ocean over time and then using this as a standard to calculate backwards to the point where the oceans would have been freshwater. Unfortunately, the salinity of the ocean is fairly constant and many factors contribute to its value, and therefore this method was never successfully employed, although as late at 1900 some scientists were still attempting to use this theory. It did represent an important first attempt at a scientific approach.

Also in the 17th century Nicolas Steno (1638–1696) developed the geological principle of superposition, which basically states that older sedimentary rocks are located deeper within the Earth and that the rock layers get progressively younger as you approach the surface. However, the mechanism by which these rock layers were deposited created a significant amount of controversy within the geological community. At the start of the 19th century there were two primary ideas. First was the theory of uniformitarianism, proposed by the English scientist James Hutton (1726–1797), which suggested that the earth's crust was dynamic and in a constant state of change. Opposing this was the theory that an ancient primitive ocean, covering the entire planet, had deposited all of the sedimentary rock layers at exactly the same time. This theory was called neptunism and was strongly defended by the late 18th century German scientist Abraham Werner (1750–1817). The debate between these ideas persisted well into the 19th century (see GEOLOGY), until the work of an exceptionally influential English geologist named Charles Lyell (1797–1875). Lyell was the founder of the idea that the study of present geological formations gives an indication of the past of the planet. He suggested that the current configuration of the Earth's surface was the result of millions of years of geological activity and that evidence of this activity was presented in the action of current geological processes such as erosion and volcanoes (see GEOLOGY). Lyell's concept of an Earth millions of years old, as presented in his book *The Principles of Geology* (1830), had a tremendous influence on the English naturalist Charles Darwin. Darwin's theory of natural selection as the process behind the evolution of organic life required a significant amount of time for the accumulation and selection of beneficial traits (see THEORY OF NATURAL SELECTION). Although Darwin made some estimates on the passage of geological time, he lacked the resources and expertise to accurately estimate the age of the earth. However, in the 19th century two approaches were developing that provided initial scientific estimates as to the earth's age.

The first of these once again involved the study of rock layers and the fossil records they contain. This type of research belongs to a field of science called paleontology, which became an important area of study in the 19th century (see PALEONTOLOGY). The presence or absence of fossils in a rock layer formed the basis of many early theories on evolutionary processes (see GEOLOGY, THEORY OF ACQUIRED CHARACTERISTICS, HUMAN EVOLUTION). But by the 1830s many geologists were recognizing that rock layers could be separated and identified by the types of fossils they contained. However, the question frequently arose as to how rapidly sedimentary rocks were formed. This process is dependent on the rate of deposition of the sedimentary material, which varies considerably. However, during the 19th century average rates of deposition were established for many sedimentary rocks, and the relationship between the thickness of these layers in certain areas of the world and the age of the Earth suggested a geological history of around 100 million years.

Estimates of the age of the Earth were not confined to the study of sedimentation rates and the fossil record. In their studies of the laws of thermodynamics and heat, William Thomson and Hermann von Helmholtz presented the possibility that the solar system was relatively young from a geological perspective. The laws of thermodynamics state that it is possible to convert energy between forms but that this energy conversion is inefficient (see LAWS OF THERMODYNAMICS). It was obvious that the Sun was converting its mass into solar energy and emitting heat in the process. Thomson proposed that this rate of heat dissipation could be used to estimate its age. Thomson recognized that the loss of heat from the Sun was to the result of a conversion of its mass but erroneously thought that this was because of a constriction of the Sun due to gravity (see THE SUN). It would not be until the 20th century that the true process of nuclear fission would be identified. Thomson theorized that the Sun could have been in existence for only about 100 million years, of which only the last 20 to 25 million could support life. Helmholtz suggested that at the current rate of energy conversion, the Sun could sustain itself for only another 10 million years (see SUN). Thomson then applied the same principles to calculating the age of the Earth. Geologists knew that the temperature increases from the surface to the core of the Earth and that the Earth had probably cooled from a molten state. Using known melting points and rates of cooling for rocks, Thomson was able to confirm his estimate that, like the Sun, the Earth was less than 100 million years old. Thomson was well recognized for his knowledge of mathematics and physics, and his estimates were accepted by many until the next century.

The key to understanding geological time does not lie in the study of thermodynamics or rates of deposition, but rather in the area of radioactivity and the rates of radioactive decay. In 1896 the French physicist Antoine-Henri Becquerel was studying x-rays when he discovered that potassium uranyl sulfate would emit particles without being directly stimulated (see ELECTROMAGNETIC SPECTRUM, ELEMENTS). What Becquerel had discovered was the first radioactive compound. Modern geologists recognize that the heat emitted from the Earth's core is to the result of the decay of radioactive material, a factor unknown to Thompson. By analyzing that rate of radioactive decay of uranium to lead, it is possible to establish geologic time to about 3.9 billion years ago. It is important to note that determining geologic time is dependent on the availability of rock layers for analysis.

Table 3
The Breakdown of Geological Time in the 19th Century. With the exception of the Tertiary era, these time frames were proposed or defined during the 19th century. The names of these time frames were frequently derived from the regions in which the rock layers were first identified.

Era	Period	Modern Boundaries (mya)[1]	Defined by (year)
Paleozoic	Cambrian	505–570	Adam Sedwick and Roderick Murchison (1835)
	Ordovician	438–505	Charles Lapworth (1879)
	Silurian	408–438	Adam Sedwick and Roderick Murchison (1835)
	Devonian	360–408	Adam Sedwick and Roderick Murchison (1840)
	Carboniferous[2]	286–360	William Coneybeare and William Phillips (1822)
	Mississippian	320–360	Alexander Winchell (1870)
	Pennsylvanian	286–320	Henry Williams (1891)
	Permian	245–286	Roderick Murchinson (1841)
Mesozoic	Triassic	208–245	Friedrich von Alberti (1834)
	Jurassic	144–208	Leopold von Buch (1839)
	Cretaceous	66–144	Omalius d'Halloy (1822)
Cenozoic	Tertiary	2–66	Giovanni Arduino (1760)
	Paleocene	58–66	Wilhelm Schimper (1874)
	Eocene	37–58	Charles Lyell (1833)
	Oligocene	24–37	August von Beyrich (1854)
	Miocene	5–24	Charles Lyell (1833)
	Pliocene	2–5	Charles Lyell (1833)
	Quaternary	0–2	Jules Desnoyers (1829)
	Pleistocene	1.8 mya to 11,000 years	Charles Lyell (1839)
	Holocene (Recent)	11,000 years to present	Charles Lyell (1839)

Notes:
1. With the exception of the Quaternary Era, all dates are reported in millions of years ago (mya) from current time.
2. The Pennsylvanian and Mississipian divisions are typically used only in North America.

These layers are subject to degradation by dynamic forces of the planet. The Earth itself is much older, with estimates placing its formation at about 5.4 billion years ago.

Even before a complete timescale was constructed, geologists and paleontologists were dividing the geological timescale into timeframes, called eras. Eras are then further divided into segments of time called periods, epochs and ages. Because similar rock layers may have been deposited at very different times during the geological record (see GEOLOGY), a method was needed to distinguish between these layers and to identify their precise location in geological time. The English geologist William Smith (1769–1839) recognized that the fossil record could be used to date rock layers, and thus he established an important link between paleontology and geology (see PALEONTOLOGY). However, the fossil record is not complete and thus was open to a wide range of interpretations. These varied interpretations manifested themselves in a number of different assignments of geological time. A series of conventions during the 19th century established the basic framework for the modern system of geological time, although variations persisted throughout the century. Table 3 lists the major era and periods of the geological record, the discoverer who is recognized with identifying the time and the modern recognized boundaries of the time period.

The establishment of geological time played an important role in not only the progress of 19th century geology but also the development of evolutionary theories in the biological sciences (see THEORY OF NATURAL SELECTION). It was the early evolutionists and their studies of problems of common descent among organic organisms that helped define the processes of paleontology and geologic time (see GEOLOGIC TIME, PALEONTOLOGY). Modern geologists rely on not only the fossil record for the establishment of time but also processes such as **laser-argon dating**, sometimes also called potassium-argon dating, as well as more detailed examinations of rates of radioactive decay. However, their work is firmly set on the foundation of geologic knowledge established in the 19th century.

Selected Bibliography

Albritton, Claude C. Jr. *The Abyss of Time: Changing Conceptions of the Earth's Antiquity after the Sixteenth Century.* San Francisco, CA: Freeman, Cooper & Company, 1980.

Eicher, Don L. *Geologic Time.* Englewood Cliffs, NJ: Prentice-Hall, 1968.

Gould, Stephen Jay. *Time's Arrow, Time's Cycle: Myth and Metaphor in the Discovery of Geological Time.* Cambridge, MA: Harvard University Press, 1987.

Greene, Mott T. *Geology in the Nineteenth Century.* Ithaca, NY: Cornell University Press, 1982.

Taton, Rene (ed.). *History of Science: Science in the Nineteenth Century.* New York: Basic Books, 1965.

Geology (1812–1883): The science of geology has it origins during the Scientific Revolution of the 17th century. The ancient Greek and Chinese cultures had expressed some interest in how the physical features of the Earth had developed over time, but it was not until the work of the 17th century geologist Nicholas Steno that a scientific approach was applied to the study of the earth sciences. Steno developed the law of superposition, which attempted to explain that older rocks were created first and as one progressed toward the surface the rock layers became progressively younger. This theory played an important role in the study of fossils, but also set the foundation for some important studies in the 19th century. The study of geology in the 19th century followed three major areas, all of which were interconnected in some manner. The first of these involved the study of fossils and the types of rock formations in which they were found. This initiated the study of paleontology, a hybrid discipline of the geological and biological sciences (see PALEONTOLOGY). The second major area of study in the 19th century was mineralogy, a branch of science that involves the classification of minerals and the identification of the geological processes by which they are formed. The last major area of focus was the study of physical geology. This included the first detailed explanations of the formation of mountain ranges and the nature of volcanoes, the classification of different forms of rock layers, and the establishment of the basis for a time frame of geological history (see GEOLOGICAL TIME).

At the start of the 19th century there were two major ideas on how the surface of the Earth was formed. The first of these was neptunism, which was founded on the work of the German geologist Abraham Werner. Werner proposed that the surface of the Earth had been covered by an ocean of water, and thus most rocks were of sedimentary origin. The neptunists were opposed by a group of geologists led by the French geologist Nicolas Desmarest (1725–1815), who in his studies of French geological formations suggested that the observed rock layers were primarily the result of volcanic activity. This was frequently called volcanism or plutonism. It should be noted that the concepts of neptunism and plutonism were also proposed centuries earlier by both the ancient Greek and Arabic scholars. However, Werner's and Desmarest's theories reflect more a scientific than theoretical study of geology and differed in many regards to the earlier ideas.

By the end of the 18th century geological expeditions and studies were being carried out across a large portion of the globe. In the process a wealth of information on the distribution of rock types was being collecting. At about the same time the ideas of neptunism and plutonism were also evolving into more modern ideas of rock formation. The first of these was the concept of uniformitarianism, first theorized by the British geologist James

Hutton in the late 18th century. Hutton recognized that the internal heat of the Earth was a prime force in geological activity, a concept accepted by the plutonists. Hutton also recognized, however, that sedimentary rocks were the result of long-term, uniform deposition of material and that this activity required a geological history for the Earth that was much longer than that accepted at the time (see GEOLOGICAL TIME). Opposing the theories of uniformitarianism was catastrophism, which stated that the formation of the Earth's surface was the result of a series of catastrophic floods, an idea supported by the neptunists and the religious community. Although a number of geologists supported this idea, the theory of catastrophism was best supported by the work of the French scientist Georges Cuvier (1769–1832). Cuvier presented his ideas on catastrophism in 1812 in an attempt to explain his studies of the fossil record (see PALEONTOLOGY). Cuvier's idea that the fossil record presented evidence of a series of catastrophic floods was widely supported by geologists who favored a young Earth and by naturalists who believed that all species were created at the same time (see THEORY OF ACQUIRED CHARACTERISTICS). In the early decades of the 19th century supporters of uniformitarianism and catastrophism were locked in an intense debate, each interpreting the geological record to support their theories.

The deadlock between these opposing camps was finally broken in 1830 when Charles Lyell published his highly influential series entitled *The Principles of Geology*. This work outlined Lyell's detailed studies of geological processes and the fossil record. The series may be summarized in Lyell's statement that "the present is the key to the past"; in other words, the geological processes of the modern world are the same forces that shaped the surface of the planet in the past. Lyell was an intense supporter of uniformitarianism and believed that natural forces such as weathering, earthquakes and volcanoes were sufficient over long periods of geological time to account for geological change. He supported his geological studies with widespread studies of marine fossils and recognized that the fossil record contained a history of these long-term changes (see PALEONTOLOGY). In developing his ideas, Lyell also supported the idea of an ancient Earth and was influential in classifying a number of the major geological periods (see GEOLOGICAL TIME). Lyell's work had a strong influence not only on geologists, but also on biologists such as Charles Darwin who were struggling to fit their developing concepts of evolutionary processes into the tight time frame suggested by the supporters of catastrophism (see THEORY OF NATURAL SELECTION). It is interesting to note that while Lyell's work favored the theory of uniformitarianism, modern geological thinking accepts that both theories probably are involved in geological processes. Catastrophism as an explanation for a young Earth is no longer recognized in the geological sciences. However, recent studies of the surface of Mars, as well as specific regions of the Earth, have indicated that periodic large-scale floods have played a role in shaping the observed geological formations. These studies have once again brought the idea of catastrophism to the forefront of geological thinking.

Another area of geology that was shaped by these ideas was the study of mountain ranges. The process by which mountain ranges are formed had occupied the minds of geologists long before the 19th century, and most early theories suggested that mountains are artifacts of the biblical Great Deluge. The neptunists and plutonists of the 18th century also developed theories of mountain formation that supported their cause. In the 19th century this debate continued with geologists focusing on specific mountain ranges such as the Alps of Europe and the Appalachian mountains of eastern North America. Some mountains, such as Mount Etna in Sicily, are obviously volcanic in nature, but geologists such as Christian Leopold von Buch (1744–1853) argued that it was not the deposition of lava that formed the mountain, but rather an uplifting of the Earth's crust because of the pressure of the sub-terranean lava. Others, such as the French scientist Jean Elie de Beaumont (1798–1874), suggested that there was a geographical pattern to mountain ranges on the planet and that this pattern was the result of the cooling of the Earth over time. Between 1830 and 1852 Elie de Beaumont developed the theory that the distribution of mountain ranges resembled the shape of a pen-tagon. He even initially suggested that these mountains were formed during four major, catastrophic, shrinking events in the Earth's past. Contrasting these theories was the work of Lyell, who believed that rock layers were ele-vated by some unseen force, where they were then weathered by natural forces. The sediment from the weathered mountains was deposited in the world's oceans, forming the basis for future rock layers.

The concept that lateral forces may play a role in the formation of moun-tain ranges began to take form in the 19th century, although most of the actual forces were not identified until the early 20th century. Although the majority of 19th century geological studies were conducted in Europe, a number of American geologists recognized that the European theories did not explain the formation of mountain ranges such as the Appalachians. William Rogers (1805–1881) and Henry Rogers (1809–1866) suggested that a lateral upheaval of rock layers may better explain the structure of the Appalachians, but the thought was not universally accepted. Around 1883 the American scientist James Hall (1811–1898) provided an idea that would serve as a transition between 19th and 20th century geology. Hall suggested that weight of large sedimentary deposits caused the crust of the Earth to collapse. This process is called a *geosyncline,* and it attracted the attention of many 20th century geologists where it played a role in the formation of theories of **plate tectonics** and the movement of continents.

One final advance for 19th century geology was the surveying of large portions of the globe and the improvement of geological maps. The first detailed geographic maps in Europe were made in Britain by William Smith in 1815. This was followed by the production of maps for a large portion of Europe and other areas of the globe. In many cases these maps were designed for industrial use, specifically for the identification of potential mining sites for valuable minerals, but they served the additional purpose of forming a large database of geological information. This greatly helped in the development of many of the major theories previously mentioned. The making of geological maps continues today. Organizations such as the United States Geological Survey (USGS), aided by satellite imagery, have constructed geological maps of the majority of the Earth's surface and most countries have geological teams assigned to mapping and identifying the resources of their country.

The 19th century represents a significant time for the geological sciences. Prior to this century the study of geology was conducted primarily by amateur naturalists. However, by the close of the 19th century there was a dedicated effort from an industrial and scientific perspective to organize detailed geologic studies. The result was a revolution in geological thinking, which had a tremendous impact not only on the sciences of geology, mineralogy and paleontology, but also on the developing theories of evolution within the biological sciences. As with other disciplines, geologists were required to invent new instruments to test their theories (see SEISMOGRAPH). In our time the science of geology has extended its studies to examinations of the Moon and planets such as Mars and Venus, where probes have mapped much of the terrain in an attempt to develop a more complete picture of the geological processes that may have shaped our planet billions of years ago. These studies are a direct continuation of the work of 19th century geologists.

Selected Bibliography

Baker, V. R. *Catastrophism and Uniformitarianism: Logical Roots and Current Relevance in Geology.* In Blundell, D. J., and A. C. Scott (eds.). *Lyell: The Past is the Key to the Present.* London, England: Geological Society Special Publications, 143 (1998): 171–182.

Gohau, Gabriel. *A History of Geology.* London, England: Rutgers University Press, 1990.

Greene, Mott T. *Geology in the Nineteenth Century.* Ithaca, NY: Cornell University Press, 1982.

Oldroyd, David R. *Thinking about the Earth: A History of Ideas in Geology.* Cambridge, MA: Harvard University Press, 1996.

Taton, Rene (ed.). *History of Science: Science in the Nineteenth Century.* New York: Basic Books, 1965.

Geometry (1796–1899): Geometry is frequently defined as the branch of mathematics that studies the dimensions and positioning of shapes (circles, triangles, etc.). Derived from the Greek words for *earth* and *measurement*, geometry historically was used in the surveying of land. Architects throughout time have also used geometrical thinking in the design and construction of buildings and temples. These uses of geometry predate the rise of the Greek civilization, but the ancient Greeks were the first to apply the processes of logical thinking to the study of geometry. One of the earliest and most recognized examples of Greek geometry was provided by the Greek mathematician Pythagoras (ca. 580 B.C.E.). Pythagoras formally defined the relationship between the hypotenuse and sides of a right triangle, although the Babylonians had demonstrated an understanding of this relationship centuries earlier. The major advances in the Greek study of geometry are attributed to the work of Euclid (ca. 330–ca. 270 B.C.E.). Euclid pioneered the use of an axiomatic system of geometry. In this system a series of acceptable true statements, called axioms, are established, which are in turn used to develop general geometric statements called **theorems**. Euclid's *Elements* (ca. 300 B.C.E.) was a compilation of Greek mathematics until his time and listed hundreds of such theorems. *Elements* formed the basis of a process of geometric thinking called Euclidian geometry that dominated mathematical investigations for centuries. In the 17th century a group of mathematicians led by René Descartes and Pierre de Fermat effectively applied the use of algebra to geometry and introduced the use of coordinates to represent geometric shapes. This field is commonly called analytical geometry. Analytical geometry played an important role in the Scientific Revolution of the 17th and 18th centuries and in the formation of a branch of mathematics called calculus.

Euclidian geometry was based on the use of a compass and ruler and thus addressed only shapes that could be constructed using these instruments. At the start of the 19th century three of Euclid's axioms had challenged mathematicians for centuries. Each of these had probably first been introduced by the Greek mathematician Hippocrates (460–377 B.C.E.) and then accepted as axioms by Euclid. The first of these stated that it was not possible to construct a square with an area equal to that of a given circle, sometimes called squaring the circle. The second problem, called duplicating the cube, demonstrates that by using only a ruler and compass it is not possible to construct a cube twice the volume of a given cube. The third problem is called trisecting the angle, and it states that it is not possible to trisect an angle into three equal parts using those same instruments. The problem was that although these axioms were accepted by mathematicians, they had not been proved mathematically. This changed significantly with the work of the French mathematician Carl Friedrich Gauss in the 19th century.

Gauss made many important contributions to science and mathematics, but his interest in geometry began in 1796, when he demonstrated that it was possible to construct a polygon with 17 equal length sides using Euclidian methods. This polygon is also called a heptadagon and it represented the first new geometric shape to be constructed using Euclidian geometry in more than 2,000 years. This discovery led Gauss to investigate whether there were conditions under which a polygon could not be constructed with equal-length sides. In 1801 he demonstrated that is was not possible to construct an equilateral (equal length sides) heptagon, a polygon with seven sides, using only a ruler and compass. He then progressed to mathematically show that the only equilateral polygons that can be constructed using this method must have the number of sides equal to a rare form of prime number. The final proof of Gauss's ideas was provided by the French mathematician Pierre Wantzel (1814–1848). In 1837 Wantzel also confirmed mathematically that the problems of trisecting the angle and doubling the cube were impossible using Euclidian methods. In 1882 the German mathematician Ferdinand von Lindemann (1852–1939) proved the impossible nature of the squaring the circle problem when he demonstrated that ratio of the circumference of any circle to its diameter, frequently called **pi**, was an endless, nonrepeating (transcendental) decimal (see NUMBERS AND NUMBER THEORY). Given this fact it was not possible to construct a square with the same area as a given circle using traditional Euclidian means.

Perhaps the greatest accomplishment in 19th century geometry was the recognition by mathematicians that alternative methods of geometrical analysis existed besides Euclidian methods. The root of the differences between Euclidian and non-Euclidian geometry is that Euclidian geometry works on a two-dimensional surface, while non-Euclidian geometry deals with higher numbers of dimension. The basis of non-Euclidian geometry is founded on the study of one of Euclid's more controversial axioms, sometimes called the fifth axiom. This axiom that states that if there is a point (p) that is not located on a line (l_1), then there is only one possible line (l_2) that can be drawn through the point (p) that is parallel to the original line (l_1). It is known that Gauss experimented with methods of proving that the fifth axiom was not needed to complete a geometrical proof, but he failed to make his work public. However, in the 19th century three other mathematicians decided to see whether it was possible to develop a geometric method that did not include the fifth axiom.

The first of these was the Russian mathematician Nikolay Lobachevsky (1792–1856). Lobachevsky recognized, as others had before him, that a geometric method has truths, or axioms, associated with it that are used to describe the method. Lobachevsky decided to see whether he could develop a non-Euclidian method that used a replacement for the fifth axiom. He decided that, instead of stating that only one possible line (l_2) can be drawn through point (p) and be parallel to the first line (l_1), he would attempt to develop a system under which any number of lines (l_n) can be drawn through point (p) parallel to (l_1). The idea was that if he reached a point in the process where his substituted axiom created a problem for the proof, but the original worked, then he would have established the validity of Euclid's fifth axiom. However, he could not find such a problem and in 1829 announced that he had developed a non-Euclidian method of geometry. A few years later (1832) the Hungarian mathematician Janos Bolyai (1802–1860) independently announced that he had reached the same conclusions as Lobachevsky. This discovery meant that it was possible for mathematicians to define the parameters of their method, which in turn greatly expanded the possibilities of analysis in both mathematics and geometry.

In 1854 the German mathematician Friedrich Reimann (1826–1866) developed a second form of non-Euclidian geometry. Unlike the **hyperbolic** method of Lobachevsky and Bolyai, which held that two lines could be parallel to line (l_1) and pass through point (p), Reimann developed a geometric method in which all of the lines intersected one another and no two lines were parallel. Reimann's method is best viewed as a form of geometry that would occur on the surface of a sphere, although Reimann worked out solutions for more than three dimensions. The multidimensional aspect of geometry was also being attempted in analytical geometry, most notably through the work of the English mathematician Arthur Cayley, whose work enabled the algebraic analysis of geometric figures in more than two dimensions.

One of the underlying principles of 19th century studies in geometry was to define what was an acceptable truth or axiom. Many of the axioms in use for geometry at the start of the 19th century had been derived from Euclid. The discovery of non-Euclidian methods demonstrated that it was possible to define unique forms of geometry by first establishing axioms that withstood mathematical scrutiny. This ability to construct what appear to be abstract mathematical systems is an important advance for the mathematical sciences, because many areas of physics and math require theoretical reasoning. However, there is an inherent need for a defined set of truths to establish the standard for all studies. One of the most common of these was developed at the end of the 19th century (1899) by the German mathematician David Hilbert (1862–1943). Hilbert's publication, entitled *Foundations of Geometry*, is considered by many to be the standard for defining axioms used in geometric thinking.

Selected Bibliography

Francis, Richard L. Did Gauss Discover That, Too? *Mathematics Teacher* 59 (1986): 288–293.

Grattan-Guiness, Ivor. *The Norton History of Mathematical Sciences: The Rainbow of Mathematics.* New York: W. W. Norton, 1997.

Katz, Victor J. *A History of Mathematics: An Introduction*, 2nd edition. Reading, MA: Addison-Wesley Educational Publishers, 1998.

Motz, Lloyd, and Jefferson Hane Weaver. *The Story of Mathematics.* New York: Plenum Press, 1993.

Germ Theory (ca. 1860–1884): The invention of the microscope in the 17th century unveiled to scientists a previously unknown world of life. Over the next two centuries there were a wide range of descriptive accounts of microscopic organisms. With the development of improved cell staining processes in the 19th century (see CELL BIOLOGY), it was becoming increasingly possible to study the internal structures of cells and the process by which they divided (see CELL DIVISION). Furthermore, the establishment of the cell theory in the 1830s enabled biologists to realize that the cell was the fundamental unit of life (see CELL THEORY). However, it took scientists some time to recognize that cells were biochemical factories that manipulated chemical reactions to form the molecules necessary for life (see ENZYMES, FERMENTATION). It was these introductory studies of biochemical reactions such as fermentation that led scientists to conclude that microscopic organisms may be the causing agent of many forms of disease, what is frequently called the germ theory of disease.

The pioneer in the development of the germ theory was the French chemist Louis Pasteur. Pasteur had become well recognized in the scientific community for his work on the structure of organic molecules (see ORGANIC CHEMISTRY). In the early 1860s Pasteur, one of the leading chemists in France, was called upon to solve a problem plaguing the wine industry. In some vats of wine the fermentation process was producing lactic acid instead of ethanol, thus souring the wine. In his investigation of the problem Pasteur discovered that different types of microbes were present in the two vats (see FERMENTATION), each with different metabolic properties. He then designed an important experiment to find the location of the source of these organisms (see Figure 21). First Pasteur heated a solution containing sugars and minerals to a point where all of the microbes present in the solution were killed. This procedure is now called pasteurization and is commonly used in the production of food products that contain bacteria naturally. The next step of the experiment was to attach an S-shaped tube to one of the flasks, while

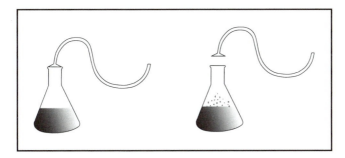

Figure 21. An illustration of Pasteur's experiments to determine whether invisible microbes in the air in France were responsible for spoiling wine. The contents of both flasks have been heated, but the left remains sealed while the right is exposed to the air. Microbial contaminants were found only in the right flask.

the other was left open to the air. The S-shaped tube inhibited possible airborne microbes from entering the pasteurized flask. After a few days the flask left open to the air was undergoing fermentation, while the flask isolated by the S-shape tube remained clear. Pasteur concluded from this experiment that the atmosphere was the source of the microbes. This discovery also played an important role in finally refuting the concept of spontaneous generation (see EMBRYOLOGY).

Pasteur made another connection between microbes and disease later in the 1860s when he was asked to assist the silk industry with a mysterious disease that was killing off large number of silkworms. Pasteur once again concluded that a microorganism was involved and suggested that leaves and silkworms currently in use by the industry be destroyed and fresh stocks brought in. This procedure eliminated the problem and stimulated additional research into the microbial relationship with diseases. Other scientists, such as Agostino Bassi (1773–1856) and Jacob Henle (1809–1885), had suggested earlier that microorganisms were associated with diseases of plants and animals, but it was Pasteur's experimental approach, reputation and interaction with important industries that promoted the concept of the germ theory within the scientific community.

The next major discovery in the development of the germ theory was provided by the German scientist Robert Koch. Koch was studying the disease anthrax, an infectious illness of cattle and other farm animals. Previous studies had indicated that this disease may be caused by bacteria and transmitted through contact between animals. In the early 1870s Koch began to study the bacteria in an attempt to understand its life cycle. To do so he invented several important instruments and procedures. The first of these was an incubator that allowed tissue and blood samples to be studied outside the organism. Next he improved the staining techniques for use with bacteria to allow a higher level of resolution while examining them under the microscope. Finally, he invented a medium made of gelatin, a semisolid material, to grow the bacteria on. Using this medium, cultivating samples and isolating individual colonies of bacteria for study was possible, an important advance in the study of bacteria. Koch's work proved that the disease anthrax was caused by a bacterium. In the process he established the basic scientific premises of determining whether a bacterium is responsible for a disease. These premises state: (1) the suspect bacteria must be isolated from every case, (2) the bacteria must be able to be grown outside the organism, (3) when injected into a test animal it must produce the same disease state and (4) the bacteria must be detected in each of these test animals. These are called Koch's postulates and are a cornerstone of microbial research. Koch went on to discover the microbial origins of tuberculosis (1882) and cholera (1884). By the end of the century little doubt remained in the scientific community about the relationship between microbes and disease. Koch was awarded the 1905 Nobel Prize in physiology or medicine for these discoveries.

With the presentation of the concept of the germ theory of disease came the development of procedures designed to limit exposure to disease-causing microbes. One of the first attempts was conducted around 1847 by Ignaz Semmelweiss (1818–1865), an Austrian physician. Semmelweiss observed that pregnant women who were admitted to the hospital for childbirth had a higher rate of puerperal fever than women who gave birth at home. When he started to study the disease in more detail he discovered that the disease also effected doctors who worked on cadavers. Semmelweiss suggested as a preventive measure that all physicians sterilize their hands using a chlorinated lime solution. As a result the rate of puerperal fever was greatly reduced in the hospitals that employed this procedure. In 1865 the English scientist Joseph Lister (1827–1912) introduced a solution of carbolic acid (also known as phenol) as an antiseptic spray. This was followed by the French scientist Casimir Davaine's (1812–1882) discovery of the antiseptic properties of iodine in 1873. Today a wide variety of antiseptic solutions are available, some of which simply use alcohol or soap, designed to limit the exposure of wounds to harmful bacteria.

There can be little doubt about the importance of the germ theory of disease to modern medicine and science. The discovery of the germ theory played an important role in the 19th century invention of vaccinations (see VACCINATIONS). The theories and procedures developed during the last decades of the 19th century are still widely used in the modern world. The pasteurization of foods, sterilization of medical equipment, and experimental procedures to isolate suspect bacteria for study were all invented in the 19th century. They represent an importance advance in science, one in which the results of the discovery benefit all of humanity and have served not only to increase our life span but also to increase the overall health of our daily lives.

Selected Bibliography

Asimov, Isaac. *Asimov's New Guide to Science*. New York: Basic Books, 1984.
Duffin, Jacalyn. *History of Medicine: A Scandalously Short Introduction*. Toronto, Canada: University of Toronto Press, 1999.
Serafini, Anthony. *The Epic History of Biology*. New York: Plenum Press, 1993.
Singer, Charles. *A History of Biology to about the Year 1900: A General Introduction to the Study of Living Things*. Ames IA: Iowa State University Press, 1989.

H

Heat (1847–1860): The modern study of heat as a form of energy originated with the Scientific Revolution of the 17th century. The work of scientists such as Robert Boyle and Isaac Newton suggested that heat was a property of matter and that it was the result of the movement of individual molecules. However, during their time the atomic basis of matter was still under development (see ATOMIC THEORY). For this reason, many scientists sought a simpler explanation for the properties of heat. By the beginning of the 19th century most scientists viewed heat as a physical substance that was contained within matter. This explanation was a common one in the early physical studies of matter. For example, until the 17th century the void between individual atoms was thought to be occupied by an element called ether. Similarly, the chemical process of combustion was thought to be the result of the release of a compound called **phlogiston**, a hypothetical, colorless and odorless compound. The existence of ether and phlogiston greatly simplified the explanation of many physical laws during the 17th and 18th centuries. However, as was the case with ether, it frequently took decades to dispel these ideas once they were established in mainstream scientific thought. A similar situation existed in the 19th century study of heat.

In the early 19th century scientists viewed heat as being a physical "fluid" called caloric. Caloric principles arose in the 18th century with the study of latent, or hidden, heat by the English chemist Joseph Black (1728–1799). Black had demonstrated that heat could be transferred between objects without changing the temperature. Antoine Lavoisier, an influential French chemist, expanded on this idea in the mid-18th century. So influential was Lavoisier's reputation that once he adopted the fluid theory of heat, it was widely accepted by the remainder of the chemical and physical scientific communities.

By the early years of the 19th century the theories of both ether and phlogiston had been disproved. At about the same time a small group of scientists were beginning to also doubt the caloric theory. Their skepticism was supported by a number of important advances in 19th century science that led to the demise of this theory. The first of these was the formulation of an atomic theory of matter by John Dalton (see ATOMIC THEORY). The atomic theory made it possible to view matter as individual atoms separated by unoccupied space. This idea led to later studies on the physical properties of gases. The development of the relationship between the temperature and volume of a gas (see GAS LAWS) and the motions of molecules within a gas (see KINETIC THEORY OF GASES) provided crucial evidence that heat was a physical, not chemical, property of a molecule. Furthermore, the study of light in the 19th century was suggesting that there were similarities between the properties of light and **radiant heat** (see LIGHT). Although there are distinct differences between radiant heat and **thermal radiation**, the relationship was sufficient to contribute to the idea that heat was not a fluid.

A number of indirect experiments also demonstrated that heat was not a physical substance. Two of these arose from 18th century observations of industry and the military. The first of these was conducted by Count Rumford (1753–1814), an American-Bavarian artillery expert who was also known as Benjamin Thompson. Rumford noticed that a significant amount of heat was produced while cannons were being bored. He speculated that if heat was indeed a fluid compound being transferred from the drill to the metal of the cannon, then the process should cease once the drill was emptied of caloric material. Furthermore, by using a dull bore, which was a slower process, less heat should be produced. Neither of these speculations proved to be true. James Watt's (1736–1819) improvements in the design of the steam engine in the late 18th century and the subsequent widespread use of steam technology in the 19th century also motivated scientist to describe the scientific basis of heat. These discoveries represent a reversal in the traditional trend of scientific discoveries resulting in technological advances for society.

However, it was not until a series of experiments were performed by the English scientist James Prescott Joule (1818–1889) that a substantial challenge to the caloric theory was developed. In a series of experiments between 1843 and 1847, Joule established what is called the mechanical equivalent of heat. In one of his experiments Joule measured the preliminary temperature of a tank of water. He then inserted a paddle into the tank and slowly turned it. Joule was able to determine that a definite relationship existed between the work done by the paddle and the subsequent rise in the temperature of the water. Over the next several years Joule worked to quantify this relationship. He eventually determined that 4.18×10^7 **ergs** of energy produce 1 **calorie** of heat. Although others had attempted to quantify this value, Joule's work represents the most accurate attempt of his time. For his contributions in this field scientists now called 10,000,000 ergs a *joule*. But of greater importance was the influence of this discovery on the study of the conservation of energy. Once Joule had established the mechanical equivalent of heat, it was possible to develop laws explaining the movement of heat, or thermodynamics. This led to the establishment of the first and second laws of thermodynamics later in the 19th century (see LAWS OF THERMODYNAMICS).

The final dismissal of the caloric theory was achieved between 1858 and 1860 when publications by James Clerk Maxwell and Ludwig Boltzmann (1844–1906) independently demonstrated that heat was directly associated with the movement of molecules (see KINETIC THEORY OF GASES). To briefly state their findings, the temperature of a gas represents the statistical average movement of the particles within the gas. When coupled to the laws of thermodynamics (see LAWS OF THERMODYNAMICS) and Joule's mechanical equivalent of heat, the idea of heat as a physical entity was effectively ended.

The study of heat represents a significant advance for 19th century science. Having entered the century with an imperfect explanation for the nature of heat, scientists worked to not only dispel the existing theory, but also to develop acceptable theories and laws based on sound experimental evidence and begin the process of quantifying energy. The kinetic theory of gases, the mechanical equivalent of heat and the laws of thermodynamics represent the results of these efforts. Each of these have had an impact on later scientists and their description of the physical properties of the universe.

Selected Bibliography

Cobb, Cathy, and Harold Goldwhite. *Creations of Fire: Chemistry's Lively History from Alchemy to the Atomic Age*. New York: Plenum Press, 1995.

Fox, Robert. *The Caloric Theory of Gases from Lavoisier to Regnault*. London, England: Oxford University Press, 1971.

Purrington, Robert D. *Physics in the Nineteenth Century*. New Brunswick, NJ: Rutgers University Press, 1997.

Spangenburg, Ray, and Diane K. Moser. *The History of Science in the Nineteenth Century*. New York: Facts on File, 1994.

Human Evolution (1856–1890): The study of evolutionary processes was a major focal point of naturalists during the 19th century. The studies of the fossil record, embryology and the geographical distribution of species on the planet were all suggesting that there was a process by which species changed over time (see EMBRYOLOGY, PALEONTOLOGY).

Added to this was the fact that by the early years of the 19th century an increasing amount of evidence was accumulating that the Earth was significantly older than claimed by the Catholic Church (see GEOLOGICAL TIME). However, the studies of evolutionary processes were typically confined to the invertebrate animals, such as **mollusks**. However, some naturalists early in the century were beginning to recognize that humans may be descended from other **primates** such as the apes. One of these was the French scientist Jean Baptiste Lamarck (1744–1829), the developer of one of the early 19th century theories of evolution (see THEORY OF ACQUIRED CHARACTERISTICS). Lamarck suggested that modern man may be a descendent of tree-living primates and that some of the modern characteristics of humans may have been shaped by environmental forces. The possibility that man may be a descendent of the lower animals was brought to the forefront of scientific thought by the work of the English naturalist Charles Darwin. Darwin's book *The Origin of Species* (1859) did not specifically apply his theory of natural selection to humans (see THEORY OF NATURAL SELECTION), but some of his later work did (see later). Many supporters of natural selection, such as T. H. Huxley, proposed that Darwin's mechanism of decent applied to all living creatures, including humans. Darwin was not the only scientist of the 19th century to state that humans may be descended from higher primates. For example, the influential English geologist Charles Lyell in 1863 published *Antiquity of Man*, which included passages on the evolution of humans. These works effectively began a controversy on the origins of humans that continues to this day.

In 1856 a discovery was made in Prussia (an area now part of Germany) that would play an important role in the developing theories of human evolution later in the century. A group of miners found the partial skull of a humanlike creature in a layer of soil that contained remains of extinct animals such as cave bears. Similar portions of skulls had been found earlier in the century, but this was the most intact specimen to date. The remnants of the skull indicated a receding forehead and protruding eyebrow ridges, characteristics most commonly associated with apes. At first some considered the abnormality of the skull structure to be the result of a bone deformity, possibly from arthritis or **rickets**. Georges Cuvier, a French paleontologist, had stated earlier in the century that no fossilized remains of humans would ever be discovered (see PALEONTOLOGY). Others thought that the remains were from an uncivilized group of humans that had earlier occupied the region. T. H. Huxley, one of Darwin's strongest supporters, recognized that the specimen was different from other human remains but would not commit to the fact that it might be a separate species. Huxley compared the skull to those of the Aborigines of Australia, a culture that was known to be more primitive than others. In his 1863 publication *Man's Place in Nature*, Huxley concluded that the skull belonged to the species *Homo sapiens*, but that it was an example of an earlier variation in the species. The first recognition by the scientific community that the skull represented a distinct species occurred in 1864. In that year the Irish scientist William King stated that the differences were significant enough to separate the group into a species called *Homo neanderthalis*, after the region in which the fossils were first discovered. The common name Neanderthal is derived from this name. More complete skeletons of Neanderthals discovered in 1886 added support for this theory. The discovery of the Neanderthal skull, plus the studies of embryology at about the same time (see EMBRYOLOGY) encouraged Charles Darwin to expand on the theme first set forth in *The Origin of Species*. In 1871 he published *The Descent of Man and Selection in Relation to Sex*, which outlined his idea that man was not exempt from the theories of evolution.

Neanderthals were not the only evidence of human evolution to surface in the 19th century. In 1868 the fossilized remains of humans were discovered in a cave in southwest France. These fossils appeared to be more than 30,000 years old, a fact later confirmed by **carbon dating** processes. These fossils belong to a group called the Cro-Magnon Man, named after the caves in which they were first discovered. In many regards Cro-Magnon man has the same characteristics as modern man and represents the earliest records of what many consider to be modern humans. In 1890 the Dutch paleontologist Marie Eugene Dubois (1858–1941) discovered the fossil remains of a humanlike creature on the island of Java in Indonesia. This fossil was more apelike than the Neanderthal and Cro-Magnon fossils unearthed earlier in the century and suggested a much deeper evolutionary history for humans. A similar fossil was discovered near Peking in 1927 by a Canadian scientist named Davidson Black (1884–1934). These fossils are now recognized as belonging to the same species, *Homo erectus*. Studies of the distribution of this species, and specifically its movement out of east Africa, have played a major role in understanding the evolution of *Homo sapiens*.

The process of carbon-14 dating has been successful for determining the age of human fossils dating back about 30,000 years, but for fossils older than that the procedure becomes inaccurate. However, a procedure called laser-argon dating, sometimes called potassium-argon dating, was developed in the mid-20th century (1964) that greatly enhances the ability of scientists to date much older specimens. Other advances in **forensic medicine** have also aided the work of paleoanthropologists. For example, a modern examination of the Neanderthal skeletons of the 19th century, the ones that led to their distinction as a separate species, indicates that the individual probably had arthritis and that it is possible that in many regards Neanderthals were much like modern *Homo sapiens* in appearance. Furthermore, they may have had a more complex social structure than originally thought. The study of these early human cultures is the basis of the field of paleoanthropology, and some members of this field now consider *Homo neanderthalis* to be a subspecies of *Homo sapiens*, a finding that has been supported by DNA evidence.

By the start of the 20th century the majority of the scientific community had accepted the concept of human evolution. However, the debate continues to this day. Although many world religions accept the idea of evolution as a biological process, they reserve the idea that the evolutionary process has been influenced by a supreme being, an idea that is now called **intelligent design**. Other groups maintain a more conservative approach, in which humans are not believed to be the result of an evolutionary process but were created in their present form, a process called **scientific creationism**. An even smaller subset of the world's **population** believes that life is a direct result of extraterrestrial interference. Evolutionary biologists, anthropologists and paleontologists may disagree as to the exact sequence of events that resulted in modern humans, but they agree that the fossil record indicates a change in our species over geologic time. Scientific creationism, intelligent design and extraterrestrial interference are not testable by available scientific methods and thus fall outside the boundaries of scientific study.

The study of human evolution in the 19th century provided some of the pieces in what was to become a much larger debate in the 20th century, both within the scientific community and between science and religion. Yet despite that fact that no resolution to the problem was really provided during the 19th century, it remains an important advance because it represents the time when scientists had finally accumulated enough resources and evidence to counter the concept of special creation. In the process they determined that the evolutionary processes that directed the changes in plants and animals also influenced humans, which in turn helped to establish a link between humans and nature.

Selected Bibliography

Darwin, Charles. *The Descent of Man and Selection in Relation to Sex*. New York: Hurst, 1874.

Mayr, Ernst. *The Growth of Biological Thought: Diversity, Evolution and Inheritance*. Cambridge, MA: Harvard University Press, 1982.

Singer, Charles. *A History of Biology to about the Year 1900: A General Introduction to the Study of Living Things*. Ames, IA: Iowa State University Press, 1989.

Tattersall, Ian, and Jeffrey H. Schwartz. *Extinct Humans*. New York: Westview Press, 2000.

Young, David. *The Discovery of Evolution*. Cambridge, England: Cambridge University Press, 1992.

I

Inheritance (ca. 1850–1900): The science of biology involves observing the living world and attempting to explain its natural patterns and phenomena in scientific terms. Since the earliest of times naturalists must have noticed that there was a regular pattern of inheritance in living organisms. From domesticated animals to humans, offspring tended to be a combination of traits from their parents. Although many consider that the analysis of these patterns began in the 19th century with the work of the Austrian monk and scientist Gregor Johann Mendel (1822–1884), in fact some explanations and experiments on inheritance date back to the time of the ancient Greeks. For example, the Greek philosopher Hippocrates, considered by many to be the father of medicine, proposed that each part of the body contributes miniature versions of itself to the formation of the embryo, meaning that the hands donate small hands, the feet contribute small feet, and so forth. This theory is called *pangenesis* and it was supported in some form until the 19th century. In fact, Charles Darwin, the developer of the theory of natural selection, adhered to this philosophy (see later). Like many theories of the ancient Greeks, pangenesis was proposed as an explanation for a general observation of nature and lacked any experimental proof to validate it as a true scientific theory. Later Greek philosophers, such as Aristotle, contested that pangenesis could not completely explain the formation of a new individual because some traits, such as behavior, lacked "parts" to contribute and the observed patterns of nature were inconsistent with a simple pangenesis model. In turn, Aristotle developed the idea that the male and female both contributed to the formation of the embryo, but in different ways. For example, the female provides the material from which the new individual will be formed, while the male provides the pattern by which the material will take shape. This mixing of functions would evolve over time into the idea that inheritance was the result of a blending of traits, a theory that had tremendous influence on naturalists until the 19th century.

Although breeding experiments continued in the agricultural industry, including the development of **pedigree** charts to signify bloodlines, the scientific investigation of the principles of inheritance went virtually unstudied until the late 18th century, when the German botanist Joseph Kölreuter (1733–1806) began his studies of plant hybridization. It had been known for some time that different species of plants could form viable offspring called hybrids. Kölreuter analyzed the hybrids of a wide variety of plant species and described the fertility and physical characteristics of their offspring. Kölreuter's work on hybridization was continued in the 19th century by a number of scientists, including Carl Friedrich von Gärtner (1772–1850) and Charles Naudin (1815–1899). By 1850 Gärtner had greatly expanded on Kölreuter's studies and had conducted more than 10,000 genetic crosses involving some 700 species of plants. In addition, Naudin's plant crosses gave an indication that a definite pattern of traits existed in the offspring of hybrid crosses. From the information derived from the work of Gärtner, Naudin, Kölreuter and others, it was possible to determine some very important principles. First, their work clearly indicated that both parents contributed information to the offspring and that there was a dominance factor among the traits. Second, there is an observed reduction in variability in the first generation of the cross (called the F_1 generation), followed by an increase in variation in the second (F_2) generation. Third, the F_2 generation frequently possessed a combination of traits of the original parental (P) generation.

The development of a modern philosophy of genetic analysis began with the work of Gregor Mendel. Mendel had broad-based training in statistics, botany (including plant-hybridization), chemistry, physics, evolutionary theory and zoology. Thus, it is likely that Mendel had an excellent working knowledge of the scientific method and experimental design, which is strongly illustrated in his work. Mendel appears to have been familiar with the work of the earlier plant hybridization scientists, and from his own observations he also concluded that there was a mathematical pattern to the characteristics of the offspring (called the **phenotype** by modern geneticists). However, unlike the earlier scientists, Mendel had the tools and desire to analyze his experimental results. Starting in 1856 Mendel began a series of genetic crosses of a common garden pea. Mendel chose the garden pea plant as his experimental system because it has physical traits that are easily identified (see later), it was easy to grow, and it was possible to self-fertilize the plants to control the crosses. Modern geneticists still choose their experimental organisms for much of the same reasons as Mendel. The results of these crosses would eventually revolutionize the biological sciences.

To begin his experimental phase, Mendel derived pure breeding lines of peas that differed slightly for seven traits (see Figure 22). The term *pure breeding* indicates that over subsequent generations these lines produced nothing but the desired trait (purple flowers, tall plants, round seeds, etc.). Mendel's first experiment involved crossing two strains of plants that varied in their physical appearance for a single trait. These variants of the trait are called **alleles**, and this type of cross is called a monohybrid cross. An example of one of these crosses is shown in Figure 23. When Mendel crossed a purple-flowered plant with a white-flowered plant, the resulting next generation (F_1) was 100% purple. The theory of blending somewhat predicted this outcome, since if one mixed purple and white dyes the resulting color would be primarily purple. However, Mendel continued the experiment by crossing two plants of the F_1 generation to derive a second generation (F_2). While ¾ of these plants had the purple flower color, ¼ had white flowers. Having a strong background in statistics, Mendel recognized the need for a large sample size, and over the next several years he repeated his experiment with each of

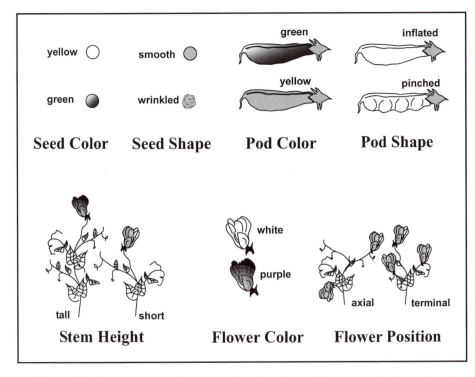

Figure 22. The seven traits of the pea plant that Gregor Mendel used in his study of inheritance.

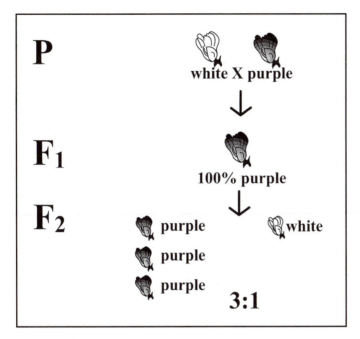

Figure 23. An example of Mendel's monohybrid cross with purple- and white-flowered plants. Mendel performed similar crosses with other traits and always obtained a 3 : 1 ratio of phenotypes (see Table 4).

his seven traits, sometimes carrying the experiment into the sixth generation to observe the ratios. In each case (see Table 4), the results indicated a clear, although not precise, 3 : 1 ratio in the F_2 generation. Mendel recognized that the minor variations from a perfect 3 : 1 ratio occurred because the production of the offspring was a random event, much like the flipping of a coin. The larger the sample size, the greater would be the chance of getting the ideal ratio of phenotypes. From this first experiment Mendel concluded not only that some traits were dominant over others (called recessive), but also that for the recessive parental trait to reappear in the F_2 generation, blending must not be occurring. Mendel also recognized that there were distinct units to the hereditary material and that to achieve a 3 : 1 ratio, each parent must possess two copies of each unit, now called **genes**. To Mendel, it was evident that these traits separate, or segregate, during reproduction, with each parent contributing a single copy to the offspring. From this Mendel derived his first law of heredity, also called the law of segregation. The modern version of this law states that during the formation of gametes in a parent, the paired alleles of a gene separate into the gametes. At fertilization, each

Table 4
The Experimental Results of Mendel's Monohybrid Crosses. The dominant
trait is listed in boldface type. The ratios are those of dominant to recessive
traits. Refer to Figure 23 for examples of traits.

Trait	F_1 generation	F_2 Generation Dominant	Recessive	Ratio
Seed color (green or **yellow**)	100% yellow	6,022	2,001	3.01:1
Seed shape (wrinkled or **smooth**)	100% smooth	5,474	1,850	2.96:1
Flower color (white or **purple**)	100% purple	705	224	3.15:1
Flower position (**axial** or terminal)	100% axial	651	207	3.14:1
Pod color (yellow or **green**)	100% green	428	152	2.82:1
Pod shape (pinched or **inflated**)	100% inflated	882	299	2.95:1
Plant height (**tall** or dwarf)	100% tall	787	277	2.84:1

parent supplies one allele, thus determining the genetic characteristics of the offspring.

This alone would have been a successful achievement for any scientist, but Mendel carried his work one additional step. Mendel was curious about whether the mathematical basis of the monohybrid cross would continue in crosses that had more than one trait. For his next experiment he developed strains of pea plants that bred true for two traits. These were crossed in what is called a dihybrid cross. An example is shown in Figure 24. As with the monohybrid cross, the individuals of the F_1 generation were crossed to obtain an F_2 generation. When Mendel analyzed the traits of the second generation, he observed that 9/16 of the offspring displayed both dominant parental traits, 3/16 possessed one of the dominant traits, 3/16 possessed the other dominant trait and 1/16 was completely recessive (see Figure 22). Furthermore, Mendel recognized that in a monohybrid cross the chances of getting two recessive alleles for the trait was 1 in 4 (1/4). Thus, if the traits were inherited independently of one another in a dihybrid cross, then the chances of getting an individual with both recessive traits would be $\frac{1}{4}$ times $\frac{1}{4}$ or 1/16, exactly the ratio that Mendel had observed. From this Mendel derived his second law of heredity, which states that the factors for different traits assort independently of one another. This is frequently called the law of independent assortment. Mendel published these results in 1866, but unfortunately they

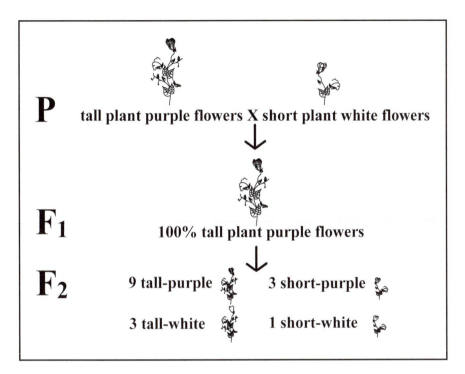

P   tall plant purple flowers X short plant white flowers

F₁   100% tall plant purple flowers

F₂   9 tall-purple   3 short-purple

3 tall-white   1 short-white

Figure 24. An example of Mendel's dihybrid (two-trait) cross. The numbers at the bottom indicate the ratio of phenotypes (9 : 3 : 3 : 1) in the second generation. This cross was important in deriving Mendel's law of independent assortment.

were published in a small regional journal and only a limited number of copies were made available. While Mendel's work effectively ended the theory of blending, established that the hereditary material moves from generation to generation as discrete particles and demonstrated that mathematical analyses could be applied to studies of inheritance, his discoveries were largely ignored until 1900 (see later).

Until the first half of the 19th century many naturalists still believed that inheritance was a blending of the fluids, or essences, of the parents, but information was slowly accumulating that there may be discrete particles involved in the transmission of traits from one generation to another. Mendel's work had suggested exactly this. In the 1880s August Weismann (1834–1914) and Karl Nägeli suggested that the genetic units of inheritance were located within the nucleus and furthermore that these units were distinct molecular objects with unique chemical characteristics. In his explanation of pangenesis Darwin had suggested that discrete particles called gemmules carried the traits to the next generation. Perhaps the most important of these ideas, however, was provided by the Dutch scientist Hugo de Vries (1848–1935). Before learning of Mendel's work, de Vries not only suggested that inheritance was the result of the independent movement of hereditary units that he called pagens, but also that it was possible for these pagens to change form and provide variation to the species. This latter statement is the basis of the theory of **mutation**, which biologists now recognize as the prime source of genetic variation.

As noted previously, the work of Gregor Mendel went relatively undiscovered by the scientific community for the remainder of the 19th century. It is possible that a few naturalists were aware of Mendel's work, but his unique mathematical approach to inheritance may have been beyond the training of most readers. However, in 1900 Mendel's work was rediscovered by three researchers: de Vries, Carl Correns (1864–1933) and Erich von Tschermak (1871–1962). Each of these individuals was working independently on the principles of inheritance when they discovered genetic ratios in their plant crosses similar to those that Mendel had discovered. When they each examined the literature to find a precedence for a mathematical link to inheritance, they discovered Mendel's previously published work. Over the next several years these researchers verified that the Mendelian ratios held true in a wide variety of plant, and later animal, species. The age of genetics had begun.

The possibility that the genetic material suggested by Nägeli and Weismann may be chromosomes was first suggested by the Oscar Hertwig (1849–1922), Eduard Strasburger and Walter Flemming as early as 1884. DNA was first detected within the cell during the 19th century (see BIOMOLECULES), but its role as the hereditary material was not completely established until the mid-20th century. Still, it was apparent from the studies of meiosis (see CELL DIVISION) that some relationship existed between the movement of the chromosomes and inheritance. In the very early years of the 20th century (ca. 1902), following the rediscovery of Mendel's work, Walter Sutton (1877–1916) and Theodore Boveri independently established that a direct relationship existed between the Mendelian ratios of inheritance and the movement of the chromosomes during meiosis. This effectively began what is now called the *chromosomal theory of inheritance*. Proof of this theory was firmly established in 1906 when the American geneticist Thomas Hunt Morgan (1866–1945) introduced the fruit fly *Drosophila melanogaster* as a model organism for genetic studies. The studies of **sex-linked traits** in this organism firmly established that the chromosomes were the carriers of the genetic material.

Questions remain to this day about the accuracy of Mendel's work, and some researchers contest that the derived ratios are almost too perfect to be the result of experiment. However, subsequent researchers have repeated

Mendel's experiments and obtained roughly the same results. Of a far greater importance, though, was Mendel's application of statistical methods to the study of genetics and his influence on early 20th century geneticists. The rediscovery of Mendel's work sparked an intense interest in the study of genetics, with scientists venturing into the fields of population genetics and the relationship between genetic variation and evolutionary theory. In the modern scientific world it is difficult to find an area of study in the biological sciences that does not study genetics and inheritance. Although the Mendelian principles that were developed in the 19th century established genetics as an important branch of science, in reality few traits adhere to the simple ratios Mendel established. Living organisms are complex combinations of genes, each one of which is susceptible to a wide range of variables, including the environment, aging, nutrition and sex, to name only a few. The development of molecular genetics techniques, specifically the production of **recombinant DNA** and **transgenic** organisms, is resulting in another revolution in the field of genetics. Projects such as the Human Genome Project have demonstrated that scientists now have a fundamental understanding of DNA. Society is now debating whether the information gained from studies of molecular genetics should be used to clone humans and develop stem cell research to cure disease. Yet, despite the development of these techniques to study genetics at the molecular level, in many cases modern geneticists are still trying to understand the principles of inheritance of specific diseases and afflictions of humans. Mendel's 19th century work provided the start to a field of study that is likely to continue for generations of future scientists.

Selected Bibliography

Bowler, Peter. *The Mendelian Revolution: The Emergence of Hereditarian Concepts in Modern Science and Society*. Baltimore: Johns Hopkins University Press,1989.

Mayr, Ernst. *The Growth of Biological Thought: Diversity, Evolution and Inheritance*. Cambridge, MA: Harvard University Press, 1982.

Serafini, Anthony. *The Epic History of Biology.* New York: Plenum Press, 1993.

Singer, Charles. *A History of Biology to about the Year 1900: A General Introduction to the Study of Living Things*. Ames, IA: Iowa State University Press, 1989.

Sturtevant, A. H. *A History of Genetics*. New York: Harper & Row, 1965.

Internal Combustion Engine (1798–1897): The invention of the internal combustion engine is commonly confused with the building of the first automobile, but the invention of the first actually predates the latter by almost a century. Since early times humans have attempted to invent methods of both easing workloads and increasing the power to do work not normally accomplishable by hand. For centuries water and animal power were the dominant methods of generating force. In the 17th century, during the time of the Scientific Revolution, the power of steam was initially harvested to drive machines and by the 19th century had become the predominant means of powering industrialized equipment (see STEAM ENGINES). Even though the 19th century was effectively the age of steam, the invention of the internal combustion engine, powered by the potential energy of fossil fuels (see FOSSIL FUELS), marked the beginning of the end of the age of steam and the start of modern industrialized society.

An internal combustion engine is defined as one in which the fuel is burned internally (usually in a cylinder) and the force of the heated gases is used to power the machine. The steam engines of the 19th century used an external fuel, usually coal or wood, to heat a container of water to form steam. One of the earliest of these machines is reported to have been first developed in 1798 by the American inventor John Stevens (1749–1838). While little is known about Stevens's machine, it is believed to have functioned by igniting alcohol in a chamber, which in turn drove a piston upward. Over the next several years Stevens made a number of improvements in the design. However, it is unclear whether this machine was used for any purpose other than demonstrations. Alcohol was not the only fuel attempted in the early part of the 19th century. Coal gas and gunpowder were also experimented with as a method of driving the pistons. One of the more interesting ideas was invented in 1820 by the Englishman William Cecil (1792–1882). Cecil used hydrogen gas as the fuel for his internal combustion engine. Cecil's machine did not gain popularity for a number of reasons, one being the relative lack of availability of hydrogen gas in the early part of the 19th century. Some inventors suggested using voltaic cells (see ELECTRIC CELL) to separate water into hydrogen and oxygen, both flammable gases, but this idea remained primarily theoretical. It is interesting to note that modern inventors are once again investigating the possibility of using hydrogen fuel cells in automobiles to reduce dependency on oil.

A large number of inventors were working on designing internal combustion engines in the mid-19th century, several which had innovative, but for various reasons unsuccessful, ideas. In 1860 a Belgian inventor named Etienne Lenoir (1822–1900) developed a prototype of the internal combustion engine that most closely resembles the modern version. Lenoir's engine was powered by gas, a fuel that was gaining popularity in the mid-19th century, although the earlier forms of these machines did not use gasoline. The first of the internal engines to use liquefied hydrocarbons (see CHEMISTRY) was probably built by the American inventor Samuel Morey (1762–1843) in the 1820s. Morey's engine used a type of carburetor to mix the liquid fuel with air before ignition. Liquid fuels such as alcohols had been used previously, but liquid hydrocarbons were more stable and would gain additional popularity as the century progressed. Lenoir's machine used a form of a spark plug that was connected to a battery. By the 1860s battery technology had progressed to the point that batteries were reliable and powerful enough to be used as an external energy source (see ELECTRIC CELL), although due to the lack of a generator Lenoir had problems keeping the batteries charged. Lenoir also pioneered the effective use of a distributor that delivered the electric power from the battery to the cylinders to allow for a sequential means of igniting the fuel. Lenoir's first machines had an estimated output of 4 **horsepower**.

Lenoir produced a number of variations of the internal combustion engine, but he is most often recognized for inspiring the work of the German inventor Nicholas Otto (1832–1891). Otto and his team of engineers are recognized as being the first to invent what is called a four-cycle engine (ca. 1876). This does not refer to the number of cylinders, but rather the sequence of events in a cycle of a single cylinder. In the first part of the sequence the piston in the cylinder compresses the fuel-air mixture at the top of the cylinder. Next, a spark is delivered to ignite the fuel and the expansion of the gases drives the piston downward, generating the force to do the work. In the third step the piston once again moves upward to expel the waste products of combustion (exhaust). Finally, the piston moves downward again, this time creating a vacuum that draws additional fuel into the cylinder. The problem with this process is that the cylinder is only working a fraction of the time, but if multiple cylinders are connected together and the ignition is alternated among them, as is the case on most modern engines, then the machine can provide a constant supply of power. From this point onward engineers in Europe and the Americas worked on improving the basic design of the multicylinder four-stroke engine. The first successful gasoline internal combustion engine is reported to have been designed by Julius Hock (ca. 1873), although evidence shows that a number of other people had earlier proposed using volatile petroleum products as fuels.

The four-cycle gas-powered engine was not the only design for an internal combustion engine that was attempted in the 19th century. One alternative form of engine was the rotary engine, which attempted to reduce the waste of force associated with the up-down movement of the four-stroke engine by attaching the cylinders in such a manner so that they rotated around a central axis. Because of engineering challenges this design was never very successful in the 19th century, although a number of manufacturers have attempted to design similar engines since then. In 1897 the German inventor Rudolf Karl Diesel (1858–1913) designed a form of internal combustion engine that did not utilize a spark to ignite the fuel, but rather the direct compression of the gas. Diesel recognized that as the volume of a gas decreased and the pressure increased, the temperature of the gas increased. He reasoned that it should be possible to use that heat as an ignition source. These machines tended to be very large, but they delivered a significantly higher power output than the spark-ignited engines. In the 19th century these engines usually used kerosene, but in the 20th century a special grade of petroleum called diesel fuel was designed to maximize their output. By the end of the century the diesel engine was being used in a number of factories in the United States.

When one thinks of the internal combustion engine today it is usually associated with the automobile. Yet most of the early engines were designed for industrial use. However, early on inventors recognized that a small engine could be used to power a carriage without the use of a horse, thus the name horseless carriage. Steam-powered horseless carriages had already been invented (see STEAM ENGINES), but they tended to be very large machines. A number of inventors did experiment with placing small internal combustion engines on chassis, but for the most part these were isolated events and not widely produced. The first production model of what can be considered an automobile was constructed by the German engineer Karl Benz (1844–1929) in 1885. It was a three-wheel vehicle with the engine mounted in the rear. The engine was small, barely 1 horsepower, but it did have the ability to move the automobile along at about 10 miles per hour. Many consider the American industrialist Henry Ford (1863–1947) to be the inventor of the automobile, but Ford's most important contribution was that he applied the process of mass production to the automobile industry at the end of the century. Ford borrowed ideas from the American firearm manufacturer Eli Whitney (1765–1825) as to the use of assembly lines and identical parts. Along with the use of workers specialized in specific tasks, Ford's idea enabled him to produce large numbers of automobiles by the end of the century.

There can be little doubt as to the influence that the internal combustion engine had on society. By the end of the 19th century the age of the horse was effectively over, and the dominance of steam-powered machines was diminishing. The internal combustion engine not only powered the later years of the Industrial Revolution, but it also transformed the Western world. What the railroad did for the mass movement of goods and raw materials, the internal combustion engine did for the movement of people. Gas engines allowed access to markets that rail or ship did not. Over the next century improvements such as fuel injection and computerized monitoring systems have increased the efficiency and power of the machines significantly. However, while the internal combustion engines of the late 19th century are primitive in comparison to the engines in use today, the fundamental principle remains effectively the same and thus is a tribute to the 19th century inventors who pioneered its use.

Selected Bibliography

Asimov, Isaac. *Asimov's New Guide to Science.* New York: Basic Books, 1984.

Hardenberg, Horst O. *The Middle Ages of the Internal Combustion Engine: 1794–1886.* Warrendale, PA: Society of Automotive Engineers, 1999.

Karwatka, Dennis. *Technology's Past, Vol. 2: More Heroes of Invention and Innovation.* Ann Arbor, MI: Prakken Publications, 1999.

Van Dulken, Stephen. *Inventing the 19th Century: 100 Inventions That Shaped the Victorian Age from Aspirin to the Zeppelin.* New York: New York University Press, 2001.

K

Kinetic Theory of Gases (1808–ca. 1870): The development of a unifying theory on the motion, or kinetics, of atomic particles was the culmination of several diverse areas of 19th century science. Although frequently regarded as a 19th century achievement, the idea that the atoms of gases are in motion was considered as early as the 17th century. In the early 18th century the Swiss mathematician Daniel Bernoulli (1700–1782) proposed a kinetic theory on the movement of gases that gave each gas atom independent movement in space. Despite the accuracy of his prediction, Bernoulli's theory contradicted the prevailing idea of the time that the space between atoms was filled with an invisible compound called ether. However, by the early decades of the 19th century many physicists and chemists were investigating the physical properties of gases. Because of the loose association of their atoms, gases were the preferred form of matter used in investigations on the atomic theory of matter in the early years of the century (see ATOMIC THEORY). This was followed by a series of studies that established some of the fundamental principles of gases. Later grouped into what are called the ideal gas laws, these principles described how the temperature, pressure and number of atoms influence the volume of a gas (see GAS LAWS). In the 1840s the definition of the nature of heat was under revision as scientists recognized that heat was a form of energy. In these investigations gases once again played a key role (see HEAT, LAWS OF THERMODYNAMICS). Each of these areas of study yielded important individual advances for 19th century science. However, by the end of the century a group of scientists were able to combine the principles of these assorted disciplines to develop a theory on how gases behave at the molecular level.

Credit for the synthesis of these ideas into a unified theory is frequently given to James Clerk Maxwell and Ludwig Boltzmann. However, as with

many areas of 19th century science, these scientists were not truly the origi-
nators of the principles that led to their discovery. One of the first breakthroughs in the development of the kinetic theory was made in determining
the true nature of heat. Modern science recognizes that heat is actually a
measure of molecular motion, but at the start of the 19th century the majority of the scientists believed in the 18th century concept of heat as a physical
substance, such as a fluid. This idea served as the basis for what was called
the *caloric theory* of heat. While fundamentally incorrect, the disproving of the
caloric theory played an important role in early 19th century science (see
HEAT, LAWS OF THERMODYNAMICS). By the 1840s there was substantial scientific evidence accumulating that heat was molecular. Much of
this data was associated with the experiments that led up to the establishment
of the laws of thermodynamics (see LAWS OF THERMODYNAMICS),
specifically the mechanical basis of heat determined by the English scientist
James Prescott Joule in 1847 (see HEAT). In addition, evidence was mounting that a form of heat called radiant heat behaved in much the same manner
as light and thus must be associated with motion at the molecular level (see
LIGHT, HEAT).

An important link between the studies of heat and gases was made earlier
in the century (1808) by the French scientist and engineer Joseph-Louis Gay-
Lussac. Gay-Lussac played an important role in early 19th century explanations of the properties of gases (see GAS LAWS). Of importance to the
development of the kinetic theory of gases was his law of combining volumes,
which established the relationship between temperature and a volume of a
gas. According to this law, at a constant pressure as the temperature increases
the volume of a gas will also increase. This can be expressed as (see GAS
LAWS for more detail):

$$PV = nRT$$

where P represents pressure, V equals volume and T is temperature. The n
term refers to the number of moles of gas and R is a gas constant.

Once it had been determined that a relationship existed between heat and
thermal expansion, it was possible to determine the velocity of a gas molecule. Around 1857 the German physicist Rudolf Clausius (1822–1888) determined that the ideal gas laws, specifically the work of Gay-Lussac and the
17th century chemist Robert Boyle, indicated that there must be minimal interaction between the molecules within a gas. Thus, the motion of an individual molecule within a gas should be in a straight line until it collides with
another molecule. In other words, no significant attractive or repulsive forces
existed between the particles that influenced their movement. Using this information Clausius determined that an individual molecule of hydrogen travels

at a velocity of approximately 2,000 kilometers per second. However, while the velocity is great, the distance traveled by a gas molecule in a given moment of time is relatively small owing to frequent collisions within the gas that redirected the molecules in a random direction.

While a significant number of other scientists were also working on similar ideas, these experiments provided the foundation for the development of the kinetic theory of gases. Around 1859 James Clerk Maxwell and Ludwig Boltzmann independently worked out the relationship between the motion of a molecule and heat. Maxwell's work on the motion of molecular particles was derived from his earlier work on the rings of Saturn, which he had determined were not solid objects as had previously been thought, but rather were independent particles kept from coalescing by the gravitational pull of Saturn (see PLANETARY ASTRONOMY). Maxwell believed that molecules in a gas behaved in a similar manner. Through a complex series of equations, Maxwell and Boltzmann independently demonstrated that temperature of a gas reflects the statistical average movement of gas molecules, and not the movement of any individual particle. The observed fact that as the temperature of a gas increases there is a corresponding increase in pressure is the result of an increase in the number of collisions between the molecules due to the increase in velocity. This relationship is sometimes called the Maxwell-Boltzmann law or relationship, but it is frequently referred to as the kinetic theory of gases.

The kinetic theory of gases was based on the statistical movement of molecules, and yet portions of it seemed to be in conflict with the second law of thermodynamics (see LAWS OF THERMODYNAMICS). In an attempt to explain the statistical movement of gas molecules in a manner that was consistent with the laws regarding thermodynamics and the conservation of energy, Maxwell created a theoretical model system. This model consisted of two rooms joined by a single doorway. Initially, both rooms contain equal number of gas molecules. However, these molecules are traveling at different velocities based on their kinetic motion. The doorway of the room allows the fast-moving molecules (elevated temperature) to pass in one direction, and the slow-moving (lower temperature) molecules to move in the other direction. The one-way nature of the system was regulated by the theoretical demon, which represented the statistical nature of the system (see LAWS OF THERMODYNAMICS for diagram). In this system, it was possible to increase the temperature of one room without expending any energy. What Maxwell had demonstrated was a process called statistical mechanics, which represented a major breakthrough for theoretical physics.

The kinetic theory of gases was an important advance for 19th century science for several reasons. First, it provided a framework for other studies on the physical properties of gases. For example, In 1865 the Austrian chemist Johann Loschmidt (1821–1895) utilized the kinetic theory to determine the number of molecules in a given volume of gas. This is now referred to as Avogadro's number (see GAS LAWS). As noted, the kinetic theory also found widespread application in the area of theoretical physics and ushered in the use of statistics to explain complex natural systems. Perhaps of greatest importance, however, is the recognition that the kinetic theory served as a template to synthesize the advances in the areas of heat, thermodynamics and gas laws that had been achieved in the early 19th century. As such, it represented one of the major scientific discoveries of the 19th century.

Selected Bibliography

Brush, Stephen G. *The Kind of Motion We Call Heat*. Amsterdam, Netherlands: North-Holland Publishing, 1976.

Krebs, Robert E. *Scientific Laws, Principles and Theories: A Reference Guide*. Westport, CT: Greenwood Press, 2001.

Purrington, Robert D. *Physics in the Nineteenth Century*. New Brunswick, NJ: Rutgers University Press, 1997.

Taton, Rene (ed.). *History of Science: Science in the Nineteenth Century*. New York: Basic Books, 1965.

L

Laws of Thermodynamics (1824–1871): By the start of the 19th century scientists had developed a number of conservation laws in an attempt to explain the physical properties of the universe. For example, in the 18th century the French chemist Antoine Lavoisier had determined that in a closed system the amount of mass remained the same regardless of any chemical or physical reactions. This was later called the law of conservation of mass and played an important role in 18th century chemistry. A similar law regarding the conservation of momentum was proposed by the English mathematician John Wallis (1616–1703) in the 17th century. However, at the start of the 19th century no definitive work had been done on the conservation of energy in a system. This was primarily because heat energy was viewed as a physical entity called caloric (see HEAT). Caloric was believed to behave like a fluid and flow between compounds to transfer heat. But by the mid-19th century the caloric theory had been disproved and in its place was the recognition that heat was a property of the motion of molecules (see HEAT, KINETIC THEORY OF GASES).

The study of energy was a popular topic of 19th century science and a large number of 19th century scientists made some contribution to the development of thermodynamics. For example, in 1847 James Prescott Joules developed the mechanical equivalent of heat (see HEAT). Not only did this demonstrate the relationship between work and heat, but it was also one of the first accurate attempts to quantify this form of energy. Around 1860 James Clerk Maxwell and Ludwig Boltzmann independently proposed that a relationship existed between temperature and movement of particles in a gas (see KINETIC THEORY OF GASES). As these ideas developed scientists began to recognize that it was possible to develop a series of laws on the conservation of energy in a system. It is these laws that serve as the basis for the study of thermodynamics, or the movement of energy.

In 1842 the German physicist Julius Robert von Mayer (1814–1878) made one of the first formal presentations to show that the amount of energy in a system is conserved in the same manner as matter. Mayer suggested that the amount of heat generated in a system must be proportional to the amount of work produced. Unfortunately, this revolutionary idea was slightly ahead of its time, and thus Mayer had considerable difficulty getting his work recognized and published. The next year the Scottish physicist William Rankine (1820–1872) presented a theory on the conservation of mechanical energy. It states that the amount of potential (stored) and kinetic (motion) energy in a body remains constant. However, credit for the development of the first law addressing the conservation of energy is given to Hermann von Helmholtz. Although a physician, in 1847 Helmholtz presented what is often considered to be one of the most influential papers of 19th century physics. From his studies of muscle contractions, Helmholtz noted that a direct relationship existed between work and heat, a similar finding to that presented by Rankine. However Helmholtz went one step further in stating that the amount of energy in nature is a constant and that energy and matter may not be created nor destroyed. Although in the 20th century the American scientist Albert Einstein's (1879–1955) theory of special relativity (1905) mathematically demonstrated that matter and energy are interchangeable, Helmholz's work was an important first statement of what is now known as the first law of thermodynamics.

The first law of thermodynamics played an important role in the study of both physics and chemistry, where it became the basis of the study of **thermochemistry**. In addition, the defining of this law helped to explain the nature of the steam engine. Although the mechanical basis of a steam engine was well understood, energy transfer was not (see HEAT, STEAM ENGINES). For example, around 1824 the French engineer Nicolas Carnot (1796–1832) correctly associated the heat of a flame to the energy of an engine. Once again, this was the fundamental principle of the first law of thermodynamics, which stated that the energy of a steam engine was being converted from potential (fuel) to kinetic (piston motion). Helmholtz's presentation of this law made a more detailed study of these principles possible (see STEAM ENGINES). Even to an untrained observer it should be evident that the transfer of energy between systems is not completely efficient because some of the energy is lost during the conversion. To the early investigators of thermodynamics, including Carnot, this was evident because a perpetual motion machine, although theoretically possible based on the first law of thermodynamics, had never successfully been constructed. The work of two scientists, William Thomson and Rudolf Clausius, further refined the principles of thermodynamics to explain this phenomenon.

William Thomson, also known as Baron Kelvin, made a number of important contributions to the study of energy. He continued the work on heat begun by James Prescott Joule and Carnot, specifically the study of the relationship between heat and work. Thomson is responsible for the development of the term *thermodynamics* (1854) to explain the flow of energy in a system and the definition of absolute zero, the point at which molecular motion ceases (see ABSOLUTE ZERO). Rudolf Clausius was a German scientist who defined a key property of thermodynamics, the concept of entropy. Entropy is the measure of disorder in a system. Clausius (ca. 1850) noted that there is always an increase in the entropy of a system. In other words, there is always a loss of energy in a reaction, and no conversion between forms is completely efficient. The combination of the work of Thomson and Clausius is called the second law of thermodynamics. This law basically states that the conversion of energy between forms is inefficient and leads to an increase in the entropy of an open system over time. The second law of thermodynamics has an important role in science in that it helps explain both the energy flow in biological systems and the physical nature of the universe.

For more than a century scientists from every discipline have been working to understand the implication of the laws of thermodynamics on the physical universe. In fact, physicists now recognize three laws of thermodynamics. The third law of thermodynamics addresses entropy at temperatures around absolute zero (see ABSOLUTE ZERO). The third law was developed by the German scientist Walther Nernst in 1906. For his work he was awarded the 1920 Nobel Prize in chemistry. The third law is an extension of a field called statistical thermodynamics, developed at the end of the 19th century by Ludwig Boltzmann. In addition, scientists have explored the theoretical conditions by which these laws of thermodynamics do not hold true. An example of this is work relates to the second law of thermodynamics, which addresses entropy and the inefficiency of energy conversion. In 1871 James Clerk Maxwell suggested that a theoretical creature small enough to manipulate individual molecules may be exempt from the second law (see Figure 25 for explanation). This creature is called Maxwell's demon, and while no such living creature exists, scientists have been working since this time to define the atomic nature of such a creature and the physical conditions that would allow it to violate the second law. These studies have important implications on the development of high-efficiency machines and has attracted theoretical quantum physicists and computer science engineers who wish to develop data-processing systems that require little or no energy. Although such a system has

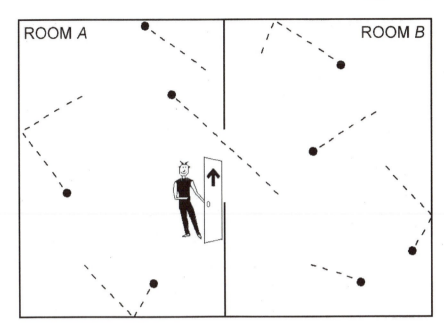

Figure 25. Maxwell's demon. This represents the theoretical concept that the second law of thermodynamics may be avoided by a creature (or object) small enough to manipulate individual molecules without expending energy. To date no such object or experimental condition has been proven to exist.

yet to be developed, the research in this area demonstrates the importance of the 19th century establishment of the laws of thermodynamics.

Selected Bibliography

Bennett, Charles, H. Demons, Engines and the Second Law. *Scientific American* (November 1987): 108–116.

Krebs, Robert E. *Scientific Laws, Principles and Theories: A Reference Guide.* Westport, CT: Greenwood Press, 2001.

Nye, Mary Jo. *Before Big Science: The Pursuit of Modern Chemistry and Physics 1800–1940.* New York: Twayne Publishers, 1996.

Purrington, Robert D. *Physics in the Nineteenth Century.* New Brunswick, NJ: Rutgers University Press, 1997.

Spangenburg, Ray, and Diane K. Moser. *The History of Science in the Nineteenth Century.* New York: Facts on File, 1994.

Light (1801–1818): Scientists of the 19th century were investigating the properties of light from many different perspectives. While some were examining the relationship between light, electricity and magnetism as part of the development of an electromagnetic theory (see ELECTROMAGNETIC THEORY), others were discovering that wavelengths of light existed outside the visible spectrum (see ELECTROMAGNETIC SPECTRUM). However, despite the progress being made in many areas of physics by the time of the 19th century, scientists entered the century still debating the true nature of light. During the 17th century, at the start of the Scientific Revolution, scientists were accumulating conflicting evidence as to whether light existed as a wave or a particle. The experiments of Francesca Grimaldi (1618–1663) and Christiaan Huygens (1629–1695) demonstrated that certain properties of light, namely diffraction, could best be explained if light was a wave. However, this was not always the case. For color and polarization, the particle nature of light was believed to be more applicable. This idea, also called the corpuscular theory, was supported by the English physicist Isaac Newton. Such was the power of Newton's reputation that many scientists of the 18th century continued to support the corpuscular theory despite the mounting evidence that light behaved like a wave.

The first real challenge to the corpuscular theory began in the early 19th century with the work of the English physician Thomas Young. At the beginning of the century physicists considered the void between atoms to be filled with a material called ether. This idea had been around in some form since the time of the ancient Greeks. The presence of ether greatly simplified the explanation of the movement of energy through a vacuum in much that same way that caloric was used to explain the process of heat transfer (see HEAT). Between 1801 and 1803 Young presented several papers to the Royal Society on the wave nature of light. In these presentations Young proposed that the movement of water through the ether was analogous to the movement of sound through air or water. In support of his ideas Young presented the results of a very simple, but conclusive experiment that was based in part on studies by Grimaldi in the 17th century. In Young's experiment (see Figure 26), a beam of light was passed through a small opening onto a paper with two openings cut into it. Young recognized that if the corpuscular theory were correct, and light was a particle, then only two individual bands should appear as the light passed through the opening. Instead, the results indicated a series of light and dark bands. Young explained that since light was a wave that behaved in much the same way as a wave on the ocean, when the waves of light from each opening overlapped, the beam was reinforced and a light band appeared. However, when the two beams were out of phase, they cancelled each other out and a dark band appeared. This phenomenon was later called the law of interference. Unfortunately, Young's work was not immediately accepted by the scientific community,

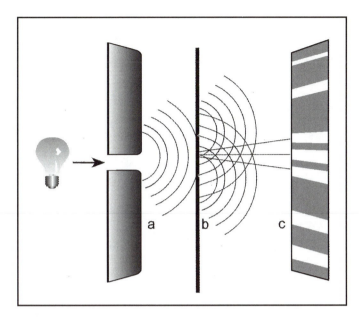

Figure 26. Young's experiments show that light has wavelike characteristics. The pattern of bands to the right, called interference bands, indicate the summing and canceling out of the light waves as they pass through the narrow slits in the center of the diagram.

because it did not at first seem to explain the processes of diffraction and polarization.

The explanation for diffraction was provided in 1818 by Augustin Fresnel, a French engineer and physicist. Diffraction is the apparent ability of light to bend around an obstacle in its path. Newton thought that this ability presented a severe challenge to the wave theory of light, because if light and sound were similar, then why was it not possible to see a person around the corner of a building when you can hear them? The answer to this question began with Grimaldi's work on diffraction in the 17th century. However, it was Fresnel who established the mathematical basis of light diffraction. In a series of complex and highly accurate mathematical calculations, Fresnel expanded on the 17th century work of Huygens and in the process demonstrated that the most reasonable explanation for the diffraction of light was if light existed as a wave. In doing so he proposed that diffraction was a process of interference between an object and a wave of light. Such was the power of his mathematical proof that by the mid-1800s few scientists still openly accepted the particle theory of light.

However, there still existed one challenge to the particle theory and that was the polarization of light. It is now recognized that light exists as a complex

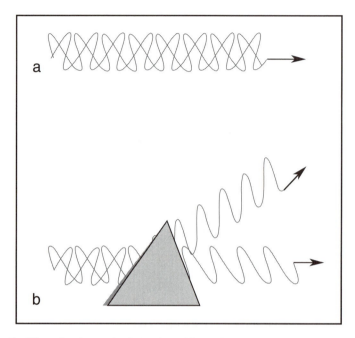

Figure 27. The refraction and polarization of light. The top diagram (*a*) indicates a beam of light. As the beam passes through an object (*b*), one of the waves is refracted upward, while one of the waves remains undisturbed by the medium. This lower wave is called a polarized wave.

series of waves. The refraction of a beam of light alters, or reflects, the path of some of these waves, while allowing only a few of the waves to pass along their original path (see Figure 27). The waves along the original path are called polarized. In 1808 the French physicist Étienne Malus (1775–1812) first used the term *polarization* to describe this process. Polarized light was an important discovery for 19th century scientists, because it allowed them to study the properties of organic compounds (see ORGANIC CHEMISTRY) and minerals (see MINEROLOGY), as well as further understand the relationship between magnetism and light (see ELECTROMAGNETIC THEORY). Polarization of light seemed to best be explained by the particle theory, with the polarized and refracted light consisting of separate particles. The synthesis of wave theory with the observed nature of polarization required the collaboration of Young, Fresnel and the French scientist Francois Arago (1786–1853), who had made important contributions to the study of electricity (see ELEC-TRICITY). Together (ca. 1818) they expanded on Fresnel's idea that light was actually a transverse wave, meaning that all points along the wave oscillate at right angles to the direction of wave movement. The explanation of light as a wave was complete and soon thereafter was accepted by most of the scientific community.

The study of light in the 19th century was not confined to the debate on wave versus particle properties. Young, Fresnel and many other scientists of the 19th century still considered the movement of light to be associated with ether, the elastic material believed to unite the universe. The study of light, specifically the relationship between magnetism and light, led to the development of the electromagnetic theory later in the century (see ELECTROMAGNETIC THEORY). Although the development of the electromagnetic theory provided experimental evidence that contradicted the existence of ether, the science of physics adhered to the belief primarily owing to a lack of an alternative theory. However, in the beginning of the 20th century, the American scientist Albert Einstein's monumental work on the physical laws of nature effectively ended support for the existence of ether and played a part in his development of the general theory of relativity. Once 19th century science had established light as a wave, it became possible to investigate the properties of color first developed by Isaac Newton in the 17th century. The result was a defining of the electromagnetic spectrum, the invention of an instrument called the spectroscope and the start of an area of science called spectral analysis (see SPECTROSCOPE). Thus the establishment of the wave nature of light played an important role in 19th century science, although it is interesting to note that the 20th century study of quantum physics once again supports the idea that light may also exist as a particle, called a photon, as part of an attempt to explain some of the complex properties of the universe.

Selected Bibliography

Bynum, W. F., E. J. Browne and Roy Porter (eds.). *Dictionary of the History of Science*. Princeton: Princeton University Press, 1981.

Park, David. *The Fire within the Eye: A Historical Essay on the Nature and Meaning of Light*. Princeton: Princeton University Press, 1997.

Ronan, Colin A. *Science: Its History and Development among the World's Cultures*. New York: Facts on File Publications, 1982.

M

Medical Instruments (1816–ca. 1860): Throughout recorded history medical practitioners have used a wide variety of devices to treat their patients. Although early physicians possessed some crude instruments, most of them were simply adapted from existing tools. Splints, knives and crude boring devices were all used in an attempt to treat patients. With the Scientific Revolution of the 17th century came some improvements in medical instrumentation. One example is the adaptation of the thermometer to measure body temperature by the Swiss physician Sanctorius. For most of recorded history, however, physicians lacked the ability to accurately assess the nature of a patient's illness. This changed in the 19th century as medical science underwent a series of important advances. After the development of the germ theory of disease and the recognition that microscopic organisms were responsible for many forms of illness (see **GERM THEORY**), scientists experimented with vaccinations and in the process greatly limited the spread of diseases such as smallpox (see **VACCINATIONS**). In addition, chemists and physicians were experimenting with methods of reducing pain through the use of anesthetics during surgery and childbirth (see **ANESTHETICS**). For this and other reasons, the 19th century is frequently viewed as the beginning of modern medicine. Along with the modernization of medical procedures developed a greater specialization of medical professionals and a subsequent need for improved instrumentation. Many new medical instruments were designed during the 19th century, and the invention of the stethoscope, hypodermic syringe and instruments to examine the structure of the eye effectively demonstrate the progress of medical instrumentation during this time.

One of the most recognized symbols of the medical profession is the stethoscope, an instrument used by physicians to listen to the internal sounds of the body, and especially the heart. The invention of the first stethoscope is credited to the French physician René Laennec (1781–1826) in 1816. Listening to the heartbeat of patients was nothing new for 19th century medicine. For centuries physicians had listened to the rhythm of their patients' hearts by placing their ears directly on the patient's chest. To assess for any internal abnormalities physicians frequently palpitated, or struck, the chest with their fingers and listed to variations in the returning sound. This is commonly called thoracic percussion. Such practices had been described since the time of the ancient Greeks. Where Laennec's approach differed was that he used a roll of paper to listen to the sound of a young woman's heart. After noticing the ability of the tube to detect the internal sounds of the patient, Laennec experimented with other materials in an attempt to enhance the effect. He finally settled on a cylinder of wood. Within a few years the medical community had adopted Laennec's stethoscope. Over the next several decades, variations of the stethoscope were constructed of various materials, including metals. The modern version of the instrument with two earpieces, called a binaural stethoscope, first appeared in the 1850s (see Figure 28).

Another advance for 19th century medicine was the invention of the hypodermic syringe. While the skin has long been recognized as a pathway to deliver medications into the body, until the 19th century this was primarily done by absorption. In some cases abrasive agents were first applied to the skin to remove the epidermis before administering a drug. Various forms of syringes had existed since the time of the ancient Romans, but the first instrument designed to deliver medication beneath the skin appeared around 1850. The invention of the hypodermic syringe probably has multiple origins, but there is evidence that the Irish physician Francis Rynd and the English doctor Alexander Wood used hypodermic syringes to deliver morphine to their patients around 1853 (see Figure 29). Regardless of its exact origins, the hypodermic syringe gave doctors another route by which to deliver medication to patients.

One additional area of medical instrumentation that received considerable attention during the 19th century was that for the diagnosis of disorders of the eye. One of these instruments was the ophthalmoscope, invented by the German scientist Hermann von Helmholtz in 1850. Helmholtz did not initially design the ophthalmoscope for medical purposes; instead, he was interested in using the instrument to detect whether the human eye had the ability to reflect incoming light. Helmholtz designed a device that directed light into the eye from an angle, where it was refracted back to the observer. Medical professionals quickly realized the importance of this instrument, and over the next several decades numerous improvements were made to

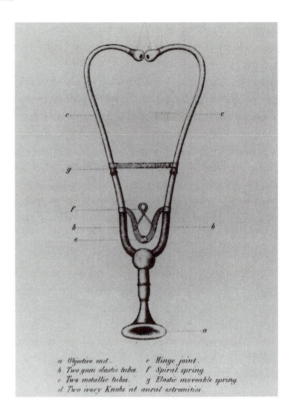

Figure 28. A 19th century stethoscope. (National Library of Medicine photo collection)

a Objective end.   e Hinge joint.
b Two gum elastic tubes.   f Spiral spring.
c Two metallic tubes.   g Elastic moveable spring.
d Two ivory Knobs at aural extremities.

Helmholtz's ophthalmoscope. The ophthalmoscope enabled physicians to examine diseases of the eye such as **glaucoma** and **cataracts**. One of the symptoms of glaucoma is increased pressure in the eye, which was frequently relieved by surgery. In the late 1850s an instrument called a ophthalmotonometer was invented that applied a light pressure to the cornea of the eye and recorded the amount that the cornea bent inward. Many consider that the practice of ophthalmology, or medicine of the eye, began with these inventions.

Despite the advances in medical instrumentation that occurred during the 19th century, physicians from that time would not even recognize the medical instruments and technology in use by today's doctor. Modern medical instruments are frequently attached to computers that accurately quantify the measurements for later analysis. Significant changes in medical instruments continue today. However, the instruments described earlier were among the first attempts by physicians to design devices specifically for the diagnosis and treatment and diseases.

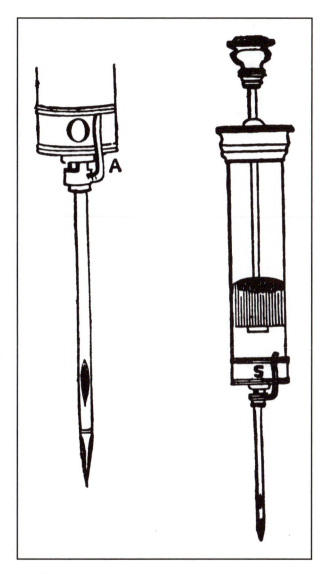

Figure 29. A 19th century hypodermic syringe. (National Library of Medicine photo collection)

Selected Bibliography

Bennion, Elisabeth. *Antique Medical Instruments*. Berkeley, CA: University of California Press, 1979.

Bud, Robert, and Deborah Jean Warner (eds.). *Instruments of Science*. New York: Garland Publishing, 1998.

Davis, Audrey B. *Medicine and Its Technology: An Introduction to the History of Medical Instrumentation*. Westport, CT: Greenwood Press, 1981.

Duke, Martin. *The Development of Medical Techniques and Treatments: From Leeches to Heart Surgery*. Madison, WI: International Universities Press, 1991.

Meteorology (1831–ca. 1895): Meteorology is the scientific study of the Earth's atmosphere, and involves research into not only the structure and dynamics of the atmosphere but also the geographical features of the planet's surface that influence the movement of air. Because the atmosphere is responsible for delivering weather systems, the science of meteorology has frequently been associated with the ability to predict the weather. However, the long-term analysis of weather patterns actually belongs to a related science called climatology. Humans have most likely attempted to predict the weather since the beginnings of civilization, but it was not until the Scientific Revolution that scientists had the means to accurately examine the atmospheric factors that influence the weather. The scientific study of meteorology, which includes the quantification of local weather information, began with a series of inventions in the 17th century. Over the course of this century scientists invented a number of instruments, including the barometer, anemometer, thermometer and hygrometers for the measurement of **barometric pressure**, wind speed, temperature and humidity, respectively. Throughout the 17th and 18th centuries these instruments became more refined and at the same time some attempts were made to examine the relationship between the atmosphere and weather at a local level. However, the atmosphere is a dynamic structure, and it is almost impossible to understand the principles by which it functions without the aid of a network of observation stations. The ability of scientists to measure widespread weather patterns, an indication of atmospheric conditions, increased significantly in the 19th century and enabled the development of new theories regarding the formation of storm systems.

Some of the earliest studies on the patterns of atmospheric movement were conducted in the 17th century by the English scientist Edmund Halley. Halley suggested that as the Earth rotates on its axis the warm air rises near the equator, only to be replaced by cooler air moving downward from the northern latitudes. Halley also proposed that this pattern of air movement was responsible for the development of the **trade winds**, a weather pattern in areas surrounding the equator. Also in the 17th century, investigators such as Robert Hooke and Blaise Pascal (1623–1662) noticed that the barometric pressure dropped as a storm approached. By the 18th century it was known that storms were associated with low-pressure systems that moved through an area, although the mechanism by which these low pressures developed and sustained themselves was not known at the time. A number of 19th century discoveries indirectly contributed information toward the 19th century understanding of how atmospheric conditions relate to storm formation. One of these was the telegraph (see TELEGRAPH), which allowed for a coordination of weather observations over an extended area. The telegraph would eventually be replaced by the radio (see RADIO), but in the mid-19th century the telegraph allowed for a system of meteorological observation stations to be established in both Europe and America. A second area of study that influenced 19th century meteorology was the development of the laws of thermodynamics (see LAWS OF THERMODYNAMICS) starting in 1847.

Another important contribution was an observation made of hurricanes by the American scientist William Redfield (1789–1857). Redfield was observing the effects of a hurricane that struck New England in the fall of 1821 when he noticed that the wind patterns associated with the hurricane indicated a counterclockwise movement of air. He conducted a number of other similar studies and published his findings in 1831. Redfield developed the theory that low-pressure systems were caused by the spinning of the air mass, which forced air from the center of the storm by **centrifugal force**, thus creating a lower area of pressure in the middle of the storm. While this may seem like a logical explanation, it took into account the measurements only of pressure and wind direction, and not temperature, associated with the storm. Another theory was proposed by the German scientist Heinrich Wilhelm Dove (1803–1879), who believed that the formation of low-pressure storm systems was due to the interaction of cold and warm air masses, which are also called the polar and equatorial air masses.

Both of these theories had some merit, but it was the thermal theory of cyclone formation first proposed by the American scientist James Pollard Espy (1785–1860) that attracted the attention of most meteorologists. Espy made an important connection between barometric pressure and temperature and how these factors related to the formation of low-pressure systems. Espy proposed that as air from the surface of the Earth is heated by the sun, it rises, bringing with it water vapor. As the column of heated air rose into the atmosphere, the barometric pressure decreased, causing the air to expand and become less dense. The water vapor in the air would then start to condense, causing the formation of a cloud. As the water condensed it would release the heat that it contained, providing the mechanism by which more air was drawn up into the storm. During Espy's time scientists still regarded heat as a physical compound called caloric, which would be physically moved in the atmosphere the same as occurred in a chemical reaction (see HEAT). Furthermore, as the air moved up the fringes of the storm would possess a greater amount of air than the core area, and thus a low-pressure center would develop. Although Espy didn't know it, later studies of the laws of heat transfer provided some support for his ideas (see LAWS OF THERMODYNAMICS). Espy's ideas were expanded upon by the American scientist Elias Loomis (1811–1889) starting in 1836. Loomis examined a large number of storm systems but chose a single storm that occurred in the United States in late 1836 as the model for his theories. From his detailed study of this system he concluded that the interaction of warm, moist air currents with cold, dry currents was responsible for the formation of storm systems. He also suggested that the apparent movement of a low-pressure system was actually due to a series of low-pressure zones that were formed along a line. The action of the rising air in the storm caused the generation of new low-pressure areas in the path of the storm, while the falling air restored higher pressure after the storm had passed. Loomis was ahead of his time since the idea of warm and cold fronts, now common features on local weather broadcasts, was still not part of meteorological studies.

Espy and Loomis had made an important connection between heat and low-pressure formation, but he could not explain why storms in the northern hemisphere rotated counterclockwise while storms in the southern hemisphere turned in a clockwise direction. In 1835 the French physicist Gaspard de Coriolis (1792–1843) provided some important information to explain the rotation of storms. Coriolis noted that since the Earth is not a perfect sphere, but rather has an equatorial bulge, the surface at the equator must travel at a higher rate of speed than the surface near the north or south poles. When an air mass moves northward or southward from the equator its speed in relation to the now slower-moving Earth beneath it increases. It the northern hemisphere this creates a counterclockwise rotation of winds, and the opposite occurs in the southern hemisphere. This is called the Coriolis effect and it had a strong influence on the American meteorologist William Ferrel (1817–1891). Ferrel recognized that it was the rotation of the Earth that interacted with the currents of air to produce a storm system. According to Ferrel, a small disturbance of pressure would be caught up by the Coriolis effect, causing the air mass to spiral. As this happened a lower area of pressure would start to form. The movement of warm air in the low pressure would provide the precipitation and the cooling of the air along the fringes of the storm provided the downward high-pressure force needed to sustain the low-pressure center. This thermal theory of cyclone formation, as proposed by Espy and Ferrel, dominated meteorological thinking for much of the 19th century. Ferrel's ideas were also used to mathematically explain why tropical storms rarely form close to the equator (where the Coriolis effect is minimal).

As the 19th century drew to a close additional information was being accumulated that suggested that the atmosphere was much more complicated than originally thought, and thus the formation of storms might not be as simple as the Espy-Ferrel model suggested. Detailed studies of clouds performed by the European and American meteorological organizations in the 1890s provided information that cloud layers were influenced by multiple layers of circulation in the atmosphere. In the last years of the century a number of meteorologists, including the Frenchman Teisserenc de Bort (1855–1913), experimented with sending meteorological instruments into the higher levels of the atmosphere using balloons. They discovered that, contrary to common belief, the temperature of the atmosphere does not uniformly decrease with altitude. At around 35,000 feet, in a layer now called the stratosphere, the temperature increases with height. Since cloud layers had been observed to occur at this level, there must be some other force at work than thermal changes.

Meteorologists now recognize that the atmosphere is actually an ocean of air, with layers, currents and disturbances in the upper atmosphere that influence the movement of air at the lower levels. The work of the 19th century scientists had indicated a connection between the movement of air in a storm and the temperature of the air. Thus the laws of thermodynamics applied to atmospheric studies. Because air in many cases also behaves as a fluid, it was possible to apply the principles of hydrodynamics as well. In 1919 Jacob Bjerknes (1897–1975), a Norwegian scientist, successfully applied thermodynamics and hydrodynamics to construct the polar front theory of cyclone formation. This theory combined the 19th century studies with early 20th century studies of the atmosphere. Briefly, it states that it is the contrast between cold air masses originating in polar regions and the warm air masses from the equator that establish low-pressure zones that form into storm systems. A version of this theory was first proposed, but not pursued, by Loomis in the 19th century. The theory has been modified since the early 20th century, but the general theory is still valid.

The 19th century was an important time for the study of meteorology, and the principles of observation and modeling of storms had a strong influence on the next generation of scientists. Modern meteorologists have a vast array of new instruments to quantify the weather, including satellites, remote sensing systems, radar and a wide network of observational stations. Instant weather information is available to a significant percent of the world's population, and modern culture demands accuracy from meteorologists. In this regard modern meteorology is vastly different from the relatively small operations of the 19th century, but the information by which forecasts are made (temperature, barometric pressure, etc.) has not changed with time.

Selected Bibliography

Bynum, W. F., E. J. Browne and Roy Porter (eds.). *Dictionary of the History of Science*. Princeton: Princeton University Press, 1981.

Kutzbach, Gisela. *The Thermal Theory of Cyclones: A History of Meteorological Thought in the Nineteenth Century*. Boston: American Meteorological Society, 1979.

Williams, James Thaxter. *The History of Weather*. Commack, NY: Nova Science Publishers, 1999.

Mineralogy (ca. 1809–ca. 1885): Minerals are nonliving chemical compounds that possess definable chemical characteristics and a uniform chemical composition. The science of mineralogy involves the classification of minerals and the study of the geological processes by which they are formed. The study of minerals had historically been of interest to commerce and industry, where the mining and trade of minerals frequently served as the basis for a significant amount of economic activity, especially in the trade of gems. As was the case with the remainder of the geological sciences (see GEOLOGICAL TIME, GEOLOGY), mineralogy underwent a significant revolution during the 19th century. During this time advances were primarily made in improving the methods of examining mineral crystals and the development of a more detailed understanding of the geological processes in the Earth's crust that led to their formation. In fact, in many ways the study of 19th century mineralogy and geology are intertwined, since many of the discoveries encompassed both sciences.

Perhaps the most significant advances in the study of 19th century mineralogy were in the development of a number of methods of classifying minerals. These classification systems included the analysis of the physical and chemical properties of mineral crystals, a process called crystallography. The study of the crystal structure of minerals was not new to the 19th century. The 17th century geologist Nicholas Steno developed a system of classifying crystals based on the geometric patterns that it formed. This was expanded on by the Swedish scientist Carolus Linnaeus (1707–1778) in the 18th century. Linnaeus is most recognized for botanical classification, a process that he adapted for use in mineralogy. Until the end of the 18th century mineralogy primarily focused on these forms of physical analysis, much of which was centered on the work of René-Juste Haüy (1743–1822), although there were a large number of other contributors. This work was expanded on in the 19th century by Gabriel Delafosse (1796–1878) and August Bravais (1811–1863). Delafosse proposed that the crystal structure of minerals was due to the symmetrical arrangement of its molecules in space. Between 1848 and 1851 Bravais developed a mechanism of studying crystals by which each molecule was treated as a single point in space, and the entire crystal structure was examined as a lattice of equally spaced molecules. By doing this he was able to classify 32 classes of mineral crystals. This discovery played an important role in the later identification of planes of symmetry within crystals.

The 19th century physical analysis of crystals did not solely focus on molecular analyses. Since the 17th century scientists had recognized that certain minerals had the ability to refract, or bend, light waves. The **double refraction** properties of a mineral called **calcite** were well studied by early investigators of optics in the 17th century, and this study continued well into the 19th century. The ability of crystals to bend light by refraction and reflection, and to generate a polarized wavelength of light, became an important tool for the studies of the nature of light by investigators such as Augustin-Jean Fresnel and Thomas Young early in the 19th century (see LIGHT). Mineralogists recognized that it was the arrangement of the molecules within the crystals that resulted in these optical phenomena, and thus polarization and refraction could be used as an additional means of identifying minerals. Drawing from crystal studies of organic materials (see ORGANIC CHEMISTRY), most notably those of the French chemist Louis Pasteur, mineralogists were able to establish the molecular configuration of some crystals and in the process develop an understanding of how the shape of the crystal relates to the observed bending of light waves. The invention of a device called a reflecting goniometer (1809) by the English scientist William Wollaston greatly enhanced the physical analysis of crystal forms. The reflecting goniometer, an improvement of a device first invented a century earlier, used a reflective brass plate to measure the angles of the surfaces on the face of a crystal. This instrument had the capability to measure the face of a crystal that was 1/50 of an inch across. For this reason many consider the reflecting goniometer to be one of the more important advances in the study of mineralogy.

Despite the increased amount of information being derived on the structure of minerals from physical analysis, it was well recognized that many minerals were polymorphic, or had multiple forms. While some minerals were polymorphic, others possessed only one (isomorphic) or two (dimorphic) forms. Sometimes very different minerals had the same crystal structure, a process called homeomorphism. Many mineralogists adopted chemical methods of analyzing these forms and in the process started to identify the elements associated with various types of minerals. Such studies were not confined solely to terrestrial objects. During the 19th century an interest developed in the chemical studies of minerals obtained from meteorites. These early studies were among the first to identify the abundance of nickel in meteoric minerals, which in turn led to the use of a classification system for meteors based on their mineral content. Not only was this important in the development of an understanding of how minerals form on Earth, but it also played an important role in the recognition that terrestrial processes of mineral formation may also have occurred at other locations with the solar system. By the close of the century mineralogists had developed a number of tools for analyzing minerals and had proposed multiple systems of classification based on physical and chemical characteristics.

Despite the work being conducted on mineral structure, and the realization that these structures were to the result of the long-term exposure of minerals to geological forces such as heat and pressure, there was still a considerable amount of debate in the early part of the century on the nature of the geological conditions that gave rise to rock formations. Rocks are collections of minerals and are frequently identified by their mineral content. In the 19th century the science of mineralogy gave rise to two additional areas of geology dedicated to the study of rocks. The first, called petrography, is primarily involved with the description of different rock types. During the 19th century petrographers agreed that there were three primary types of rocks: igneous, metamorphic and sedimentary (see GEOLOGY). A more detailed examination is performed by geologists specializing in petrology, which examines the forces responsible for rock formation. Petrologists are more likely to include chemical analysis in their descriptions of rocks. In the 19th century petrologists were the first to use microscopic examinations of mineral crystals to attempt to explain rock formation.

Modern mineralogists rely on many of the methods of classification developed during the 19th century. However, there have been additional advances, mostly in the ability to perform precise chemical analysis using the process of mass spectroscopy. The physical structures of crystals have further been determined through the discovery in the early 20th century of x-rays (see ELECTROMAGNETIC SPECTRUM). This has enabled the development of an exceptionally useful procedure called x-ray crystallography, which has found application in a wide variety of scientific disciplines. The 19th century is frequently viewed as the starting point for the scientific study of mineralogy, which is a result of the dedicated interest of the geological community to explain this important aspect of their discipline.

Selected Bibliography

Bud, Robert, and Deborah Jean Warner (eds.). *Instruments of Science*. New York: Garland Publishing, 1998.

Laudan, Rachel. *From Mineralogy to Geology: Foundations of a Science 1650–1830*. Chicago: The University of Chicago Press, 1987.

Oldroyd, David R. *Sciences of the Earth: Studies in the History of Mineralogy and Geology*. Algershot, UK: Ashgate Publishing, 1998.

Taton, Rene (ed.). *History of Science: Science in the Nineteenth Century*. New York: Basic Books, 1965.

N

Nitroglycerine (1845–1868): Nitroglycerine belongs to a class of chemical compounds frequently labeled as high explosives. The invention of nitroglycerine in the 19th century represents the first of these compounds to be discovered. However, the use of various forms of explosive chemicals is recorded by ancient Greek and Arabic cultures. The first explosive, and not simply an incendiary device, was gunpowder. Gunpowder is a combination of charcoal, sulfur and potassium nitrate, commonly called saltpeter. The purpose of the saltpeter is to provide the oxygen needed for the combustion of the other materials. There is some debate as to the first culture to develop gunpowder, but many historians agree that it was probably the Chinese in the 9th century. The English philosopher Roger Bacon (1214–1292) introduced gunpowder to the Western world in 1260, He most likely learned of gunpowder from Arabic sources, which frequently served as an important conduit of knowledge from Asian to European cultures during this time. Gunpowder is a widely known explosive compound, but it has a relatively weak energy output when compared to the high explosives of the late 19th and the 20th centuries. Furthermore, it is almost completely unusable when wet. For most of its history it found application in launching small projectiles, such as fireworks, or in the military.

In 1845 the German chemist Christian Shonbein (1799–1868) accidentally discovered that the combination of nitric and sulfuric acids produced a substance that reacted violently to heat. When mixed with a cellulose material the compound is called nitrocellulose, although it is more frequently known by its common name, guncotton. A few years later (1847) the Italian chemist Ascanio Sobrero (1812–1888) invented nitroglycerine, which is sometimes also called by its correct chemical name, glyceryl trinitrate (see Figure 30). Sobrero's invention was not initially intended as a high explosive. Sobrero had

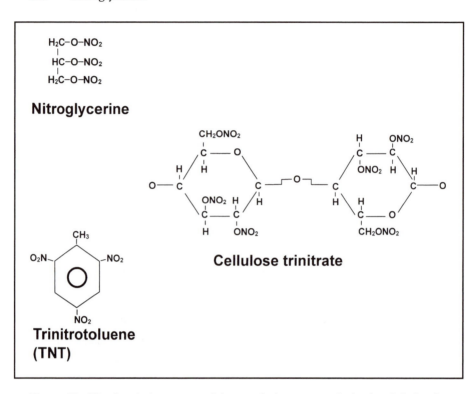

$H_2C-O-NO_2$
$HC-O-NO_2$
$H_2C-O-NO_2$

Nitroglycerine

Cellulose trinitrate

Trinitrotoluene (TNT)

Figure 30. The chemical structures of three explosive compounds developed during the 19th century.

a background in medicine and some of his research involved investigating chemical means of reducing chest pain, commonly called angina. In one of his experiments Sobrero combined **glycerine** with nitric and sulfuric acids to produce a yellow oil. The term *nitroglycerine* is derived from the chemical components of the oil. Unfortunately, tests of this compounds on laboratory animals indicated that it had some rather unpleasant side effects, namely a rapid heart rate, vomiting and body tremors. It also was a highly unstable compound, and Sobrero documented several explosions in his laboratory, some of which were simply caused by vibration. Over the next decade a number of other doctors explored the use of smaller amounts of nitroglycerine to treat chest pains and headaches. By the late 1850s it was discovered that nitroglycerine could effectively be used to treat these medical problems. It is now recognized that nitroglycerine acts by dilating the blood vessels, a fact first noticed by Sobrero in his studies. It is now widely used for patients with heart problems.

The first investigations of nitroglycerine as an explosive began in the 1860s by Alfred Nobel (1833–1896). Nobel's father was an inventor who also specialized in the design and trade of military equipment. However, he also had an interest in developing explosives to be used in civil engineering projects such as road and canal building. Not only were gunpowder and nitrocellulose somewhat unstable, but also they lacked the power to blast through the rock layers for these projects. Nitroglycerine was also unstable but much more powerful. The first task was to stabilize the nitroglycerine so that it could be detonated when desired by the engineers. In 1863 Alfred Nobel invented a detonator that was made from a chemical compound called mercury fulminate. Prior to this detonators used either gunpowder or guncotton to ignite the nitroglycerine, but the use of mercury fulminate in the detonator provided a more reliable firing mechanism. Despite the use of detonators, the control of a nitroglycerine reaction remained unreliable, as was evident by an explosion at the Nobel's factory in Germany in 1864 that killed five people, including Alfred Nobel's brother.

In an attempt to further control the explosive properties of nitroglycerine, Nobel focused on transforming the compound from its first discovered liquid state to a more stable solid state. To do this he added a chalklike compound called diatomite. Diatomite, also called diatomaceous earth, is the fossilized remains of small microscopic creatures called **diatoms** and contains a high amount of the mineral calcium. Diatomite has the ability to absorb a large quantity of nitroglycerine and still remain in a solid state. The combination of diatomite and nitroglycerine produces dynamite, which Nobel derived from the word *dynamic*. Unlike nitroglycerine, dynamite has a much higher degree of stability and is less susceptible to accidental detonation by vibration, as was the case with pure nitroglycerine. However, dynamite is not entirely stable and has been known to deteriorate over time. During the 19th century dynamite was detonated using blasting caps and fuses, although there are now electronic circuits that can be used to precisely control the timing of the blast. In 1867 Nobel received a patent for dynamite in England, followed the next year with a patent in the United States. Dynamite quickly became the explosive of choice not only for civil engineers but also the military.

Perhaps one of the greatest contributions of the discovery of nitroglycerine to the scientific world has nothing to do with the explosive or medical properties of the chemical. Through the sale of dynamite and nitroglycerine Alfred Nobel had prospered to millionaire status. Although he was well known in some circles for his philanthropy, in the later years of his life he became dismayed that the media continuously reported him as the "dynamite king" and the "merchant of death." Determined to leave something beneficial to mankind other than explosives, Nobel set up a foundation to award achievements in several fields. Although the basis for these awards was fiercely contested by the Swedish government and members of his family after his death, the Nobel prizes for Literature, Chemistry, Physics, Peace, Economics, and Physiology and Medicine have been awarded to international recipients since 1901.

During the 19th century other high-explosive chemicals were introduced besides nitroglycerine. One of the better known of these is trinitrotoluene (TNT), which was invented in 1841 (see Figure 30), although because of production difficulties it did not see widespread use as an explosive until just prior to World War I. By the end of the century many industrialized nations across the globe were experimenting with improved high-explosive chemicals and the methods of manufacturing them in large quantities. Many of the great naval battles of World War I were won owing to the availability of these explosives. However, during the 19th century it was nitroglycerine that dominated the market, and it remains one of the most popular explosives in civil engineering to this day. When coupled with the medical importance of the chemical, nitroglycerine may be regarded as one of the more important scientific breakthroughs in chemistry during the 19th century.

Selected Bibliography

Brown, G. I. *The Big Bang: A History of Explosives*. Phoenix, AZ: Sutton Publishing, 1998.

Karwatka, Dennis. *Technology's Past, Vol. 2: More Heroes of Invention and Innovention*. Ann Arbor, MI: Prakken Publications, 1999.

Van Dulken, Stephen. *Inventing the 19th Century: 100 Inventions That Shaped the Victorian Age from Aspirin to the Zeppelin*. New York: New York University Press, 2001.

Numbers and Number Theory (1801–1882): In mathematics the study of number theory involves defining the properties of whole integers. It is an ancient branch of mathematics, with evidence suggesting that the studies originated as far back as the 6th century B.C.E. by Pythagoras. The Greek mathematician Euclid compiled one of the first comprehensive studies of ancient number theories, which he published as part of his book *Elements* in the 3rd century B.C.E. Over the next several centuries various Roman scholars studied number theory, and during the Dark Ages of Europe the knowledge of number theory was kept alive by Arabic mathematicians. There was a revival of studies in number theory in Europe during the 17th century, most notably by the French mathematician Pierre de Fermat. Fermat was interested in the properties of prime numbers, those numbers that have only 1 and themselves as factors. Perhaps Fermat's most important contribution to the history of number theory is Fermat's last theorem, a definition of conditions under which certain positive integers can't exist. In the 19th century a large number of mathematicians were working on some aspect of number theory as it related to the study of algebra and geometry (see ALGEBRA, GEOMETRY). However, there were several significant breakthroughs in the study of number theory that played an important role in the development of 20th century mathematics.

One of the first important advances in the century was proposed by the influential German mathematician Carl Friedrich Gauss. In 1801, at the age of 21, Gauss published a book entitled *Investigations in Arithmetic*, which contained several of his contributions to number theory. One of these was the theory of congruences, which uses the format of algebraic equations to define the relationship between numbers, and specifically prime numbers. This work enabled mathematicians to quickly analyze large numbers to determine their factors and whether they could be classified as prime numbers. The study of prime numbers was a major focus in 19th century number theory research. The French mathematician Adrian-Marie Legendre (1752–1833) worked out a system of estimating the density of prime numbers in a given range of values. While at first it may seem that prime numbers should be relatively evenly distributed, there are actually areas of dense prime numbers followed by areas of lower density. Only when one looks at large expanses of numbers does there appear to be a mathematical basis. This feature of prime numbers was also observed by Gauss.

Other 19th century mathematicians similarly concentrated on defining the properties of specific types of numbers. One of these classes is the transcendental numbers. The word *transcendental* is derived from the Latin for "to climb beyond" and was first used by Leonhard Euler (1707–1783), a leading 18th century Swiss investigator of number theory, to describe a group of numbers that could never be defined by an algebraic equation. This means that they have an infinite number of nonrepeating decimal places, or are **irrational**. The first of these transcendental numbers to be identified as being unobtainable using algebraic means is called e. The quantity e, which is approximately equal to $2.71828 \ldots$, was first hinted at in the 17th century during the study of **natural logarithms** by mathematicians such as Christiaan Huygens and Nicolaus Mercator (1620–1697). Jacob Bernoulli (1654–1705) gave the first approximation of the value of e during his study of compound interest in the late 17th century. He recognized that it was associated with the equation $(1 + 1/n)^n$ as n approaches infinity. Several years later the German mathematician Gottfried von Leibniz, cofounder of calculus, formally identified the number and in the 18th century Euler provided the current notation of e and proved that it was irrational. However, the proof that e was transcendental was first provided in the 19th century (ca. 1873) by Charles Hermite

(1822–1901), a French mathematician who made several important contributions to 19th century mathematics. The number e has been demonstrated to be irrational and transcendental through the use of modern computers and continues to play an important role in the study of engineering and physics.

Another of the more recognized irrational numbers that was studied in the 19th century is pi. Pi is an ancient number that has been recognized for several millennia to represent the ratio between the circumference of a circle and its diameter. The ancient Greek mathematician Archimedes (ca. 287–212 B.C.E.) is considered by many to have developed the first recorded proof of this value. For centuries afterward Chinese, Arabic and European scholars worked on calculating the number of decimal places in pi and developing formulas to approximate its value. The fact that pi is an irrational number was first proved by the Swiss mathematician Johann Lambert (1728–1777) in 1761. In 1882 pi was proved to also be transcendental by the German mathematician Ferdinand Lindemann, who used Hermite's work on e as the basis for his proof. Lindemann's proof effectively proved that the ancient problem of squaring the circle (see GEOMETRY) was impossible.

The 19th century study of numbers and number theory, when coupled with the explorations in geometry and algebra (see ALGEBRA, GEOMETRY), mark what many consider to be the golden age of mathematics. Mathematicians during this time were learning to redefine the principles of mathematical analysis and in the process developed theoretical and abstract methods of analyzing previously unrecognized realms. This type of thinking would be important in developing the theories of relativity, quantum physics and nuclear fission, among others, during the 20th century.

Selected Bibliography

Katz, Victor J. *A History of Mathematics: An Introduction*, 2nd edition. Reading, MA: Addison-Wesley Educational Publishers, 1998.

Kline, Morris. *Mathematical Thought from Ancient to Modern Times*. New York: Oxford University Press, 1972.

Motz, Lloyd, and Jefferson Hane Weaver. *The Story of Mathematics*. New York: Plenum Press, 1993.

O'Conner, John J., and Edmund F. Robertson. 2002. *The MacTutor History of Mathematics Archive* (http://www-groups.dcs.st-and.ac.uk/~history/).

O

Organic Chemistry (1823–1887): By the time of the 19th century the science of chemistry had already been divided into two general categories. The science of physical, or inorganic, chemistry involves examining the structure of elements, molecules and compounds and how they interact by chemical processes. The more specialized branch of organic chemistry comprises the study of the chemical reactions associated with the element carbon. Chemistry as a science has its roots in the 17th century, but the study of organic chemistry can be said to have developed in the 19th century and represents one of the greatest scientific achievements of this time.

Carbon is the fundamental element of all living things on this planet, and for this reason the study of organic chemistry is frequently tied to the study of organic biological processes. This relationship between carbon and life was well recognized, but poorly understood, in the early 19th century. However, as the century progressed advancements were being made in several areas. In 1823 the German chemist Friedrich Wöhler (1800–1882) was studying the composition of cyanic acid, a compound that contains a single atom of nitrogen, oxygen, silver and carbon. At about the same time, another German chemist, Justus von Liebig, was performing a similar investigation of fulminic acid. Liebig determined that this compound contained a single atom of these same elements. By coincidence, both men published their results to the same scientific journal. This created a significant amount of confusion since fulminic acid and cyanic acid are chemically different but have the same molecular formula. Liebig and Wöhler had discovered *isomers*, compounds with similar composition but different chemical properties. While this made it clear that the arrangement of atoms within a molecule determined chemical reactivity, it was unclear how carbon was involved in the formation of these different structures.

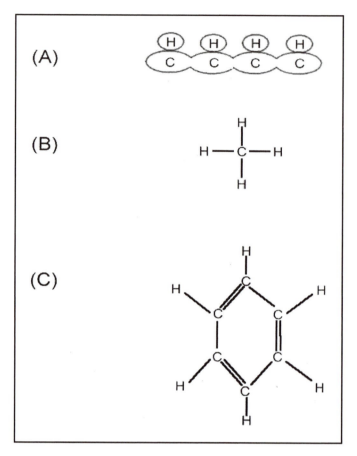

Figure 31. Three studies of organic chemistry in the 19th century. The top diagram (*A*) represents Kekulé's diagram of tetravalent carbon, and the second diagram (*B*) illustrates Brown's accepted portrayal of carbon. The lower diagram (*C*) illustrates the ring structure of benzene, a compound that helped explain the tetravalent properties of carbon.

The next major advance in the study of organic molecules was the concept of valency. The valence of an element or compound refers to its combining power in a chemical reaction. This idea arose from the work of a number of chemists, primarily Charles Gerhardt and Edward Frankland (see CHEMISTRY). By the mid-19th century chemists had determined that all elements have a characteristic valence and that these variations could be utilized in the construction of a periodic table for the elements (see ELEMENTS). In 1858, the Scottish scientist Archibald Couper (1831–1892) determined that carbon was a tetravalent element, meaning that each carbon atom had the ability to combine with four other elements (see Figure 31). This discovery is frequently credited to Friedrich Kekulé, who was the first to publish these results in an accredited journal. The tetravalent structure of carbon made it a versatile element, and with this discovery began the study of chemical structure. Kekulé suggested that the carbon atom could be responsible for the formation of a skeleton for the organic molecule. The idea that organic molecules may not just be a loose association of atoms, but may actually have an established structure, had begun. The idea of a carbon backbone for all organic molecules was formally presented by the Russian chemist Aleksandr Butlerov (1828–1886) around 1857.

With the concept of structure arose the question of how to illustrate these structures. The first chemical structures drawn by Kekulé portrayed the individual atoms as vague clouds. The overlapping of these atomic clouds represented the bonding between the atoms (see Figure 31). A more simplistic approach was proposed by Couper, who suggested that dotted lines represent the interaction between the atoms. This idea was expanded upon in 1864 by the English chemist Alexander Brown (1838–1922), who formalized the method of drawing chemical structures that is still primarily used today (see Figure 31). However, it should be noted that although chemical bonds are drawn as lines in almost all chemical publications, in reality it is now recognized that chemical bonds are formed by the interaction of electron clouds around the atom, a tribute to Kekulé's diagrams constructed decades before the discovery of the electron (1898).

These advances allowed chemists to begin to explore the possible structures of organic molecules. For example, the compound benzene was first isolated in 1825 by the English chemist Michael Farraday. However, its chemical formula indicated that it consisted of six atoms each of carbon and hydrogen (C_6H_6). This presented a problem for supporters of the tetravalent structure of carbon, because no linear molecule could be constructed that recognized the valence structure of carbon. In 1865, Kekulé proposed that a solution for this problem was to portray the molecule as a ring with several carbon molecules double-bonded to one another (see Figure 31). Since benzene compounds have a pleasing aroma Kekulé called these compounds aromatics. Ring compounds are common in organic chemistry, and Kekulé's solution proved to be an important development in the study of organic compounds.

Despite these significant advances, the study of organic chemistry was still limited by the fact that the early structural chemists considered these molecules to be two-dimensional. The three-dimensional structure of organic compounds was first proposed by the French scientist Louis Pasteur. Well recognized for his contributions to the study of microbiology, especially the process of pasteurization (see FERMENTATION), Pasteur was also trained as a chemist. He had an interest in studying how organic molecules form crystals. It was well known that many crystals were *isomorphic*, meaning that they had multiple similar structures. When Pasteur studied the crystal structure of tartaric acid, a chemical found in grapes, he noted that there were two different forms of the crystal. These forms were mirror images of one another, much like the left and right hands of the human body. He even proposed that there may be a preference for these forms, which he called *levo* and *dextro*, in biological systems. By the late 1860s many chemists were recognizing that a three-dimensional carbon molecule would best explain isomorphic compounds and the formation of double and triple bonds between carbon atoms. The **tetrahedral** structure of carbon was independently presented in 1874 by Jacobus van't Hoff (1852–1911) and Joseph-Achille Le Bel (1847–1930). Furthermore, van't Hoff suggested that two carbon atoms were able to rotate freely around a single bond, giving an even greater range of three-dimensional structures for organic molecules. Much of the three-dimensional nature of chemistry was summarized in van't Hoff's 1887 publication, *Chemistry of Space*.

The study of organic compounds in the 19th century marks one of the more important developments in the history of chemistry. The discoveries of the nature of organic molecules and their three-dimensional shape led to studies of carbon compounds in living molecules, the biomolecules (see BIOMOLECULES). Furthermore, organic chemistry quickly found application in industry where it became possible in the textile industry to manufacture synthetic dyes, increasing the durability of cloth and reducing the reliance on natural coloring agents (see SYNTHETIC DYES). The science of organic chemistry remains an important contributor to our modern culture. Many pharmaceutical products, including most prescription drugs, are synthetic organic molecules. In addition, common products, such as the artificial sweeteners aspartame and saccharin, are derived from organic molecules. Few areas of life in modern society are not influenced by organic chemistry, a science that has its origins with the early chemists of the 19th century.

Selected Bibliography

Cobb, Cathy, and Harold Goldwhite. *Creations of Fire: Chemistry's Lively History from Alchemy to the Atomic Age.* New York: Plenum Press, 1995.

Fruton, Joseph H. *Molecules and Life: Historical Essays on the Interplay of Chemistry and Biology.* New York: John Wiley & Sons, 1972.

Nye, Mary Jo. *Before Big Science: The Pursuit of Modern Chemistry and Physics 1800–1940.* New York: Twayne Publishers, 1996.

P

Paleontology (1801–1861): Paleontology is the scientific study of the fossil record, and thus represents a link between the sciences of geology and biology. Although the term *paleontology* originated in the 19th century, the scientific community has held an interest in the nature of fossils since ancient times. As early as the 3rd century Chinese scholars widely recognized that fossils represented the remains of once living creatures, an idea that also had been debated even earlier by some Greek scholars. While evidence of fossils may be found across the globe, the Western scientific community did not devote a significant amount of time to their study until the work of a few pioneers in the 17th century. Nicholas Steno is considered by many to have initiated an interest in fossils in Europe when he suggested that the fossil record could be used to explain the formation of rock layers (see GEOLOGY) and that fossilized remains often closely resembled those from living specimens. By the end of the 17th century and into the 18th century many scientists considered fossils to represent the remains of extinct species, although the time and mechanism of their extinction was unclear. During the 19th century some leading geologists, naturalists and evolutionary biologists were working on deciphering the meaning of the fossil record. In the process these studies not only would result in the formation of paleontology as a science, but also would provide some important insights into the development of 19th century theories on geological time and evolution (see THEORY OF ACQUIRED CHARACTERISTICS, THEORY OF NATURAL SELECTION, GEOLOGY, GEOLOGICAL TIME).

In the early years of the 19th century the English geologist William Smith suggested that a connection could be made between the different rock layers, called strata, and the types of fossils that the layer contained. In his 1816 publication *Strata Identified by Organized Fossils*, Smith suggested that these fossils could be used to establish a chronological history of rock formation, a process sometimes called stratigraphic paleontology. Not only did the fossil record prove to be useful to geologists, but also it allowed naturalists to construct a history of life on Earth. Although many of these early ideas were frequently inaccurate, they did promote the idea that life had progressed over time. Since the process of rock formation and geological time were not well understood at the beginning of the 19th century (see GEOLOGY, GEOLOGICAL TIME), the rate of progression and the mechanism of change remained controversial.

One of the first true paleontologists of the 19th century was the French scientist Georges Cuvier. Cuvier made a number of important contributions to 19th century science, including the development of a classification system for animals, theories disagreeing with the evolution of organic organisms and contributions to the establishment of boundaries for geologic eras (see GEOLOGICAL TIME). Cuvier combined his work in each of these areas with detailed examinations of the fossil record. One of the main problems with the fossil record is that complete, intact fossils of organisms are frequently difficult to obtain. Thus, paleontologists are constantly required to theorize on the structure of an animal or plant based on a few fossil remains, much like the construction of a puzzle without the benefit of the box top. Cuvier was one of the first to recognize that form follows function. In other words, the physical structure of an organism's anatomy is the result of what function it is required to perform for the organism. This information, in turn, could be used to determine the structure of other parts of the organism. This is called the principle of correlation. For example, flying organisms required wings plus structures such as breastbones to attach flight muscles to. Cuvier used this theory of comparative anatomy to reconstruct fossilized organisms when only a small portion of the remains were recovered, a system widely used by modern paleontologists. In 1801 he was successful in determining that the fossilized remains of a pterodactyl, or winged lizard, discovered in the 18th century actually belonged to a species of reptile. In fact, Cuvier is credited with developing a classification system for fossils that paralleled the one used for living creatures.

Although Cuvier recognized that the fossil record changed, or progressed over time, he did not support the concept that organic organisms had evolved into their current forms. Instead, Cuvier believed that all organisms had been created together and that a series of catastrophic events had eliminated large numbers of these species over time. The survivors of these mass extinctions had then repopulated the planet. He suggested that there may have been multiple creation events, each serving to increase the diversity of life on the planet. Cuvier proposed that the fossil record provided a history of these events, of which the biblical flood was one example. Furthermore, Cuvier recognized that the fossil record was not complete and believed that additional explorations would eventually prove his theories. Cuvier's theory of catastrophism, as published in 1812 as the book *Inquiry into Fossil Remains*, remained a powerful force not only in early 19th century discussions on evolution (see THEORY OF ACQUIRED CHARACTERISTICS), but also in discussions on the formation of rock layers, where he opposed the uniformitarianism theories of James Hutton (see GEOLOGY). While catastrophism as the underlying cause of evolution and geological processes was eventually disproved by mid-19th century naturalists and geologists, Cuvier's contributions to the science of paleontology marked an important turning point in the birth of this 19th century science. It is interesting to note that modern evolutionary theories have once again returned to the concept of catastrophism to explain mass extinctions in life's history.

While geologists such as Charles Lyell and naturalists such as Erasmus Darwin (1731–1802), Jean Baptiste Lamarck and Charles Darwin were developing theories that opposed catastrophism (see THEORY OF ACQUIRED CHARACTERISTICS, THEORY OF NATURAL SELECTION), paleontologists were working to establish a time frame of life's history. Paleontologists recognized that the fossil record provided a chronology of life's history. In the case of the vertebrates, there appeared to be distinct periods of time in which certain vertebrate groups dominated the Earth. Paleontologists now recognize distinct "ages" of vertebrate groups, including mammals, reptiles, amphibians and fishes (see Figure 32). The progression of vertebrate life on the planet is well documented. For example, in 1861 the discovery of a feathered reptile named *Archeopteryx* indicated a definite link between the reptiles and birds. The discovery of this fossil not only provided evidence of progression, but it also served to support Charles Darwin's theory of natural selection provided a few years earlier (see THEORY OF NATURAL SELECTION).

The study of paleontology was not limited to vertebrates in the 19th century. Because invertebrate animals represent more than 99% of all living species, they should dominate the fossil record. Unfortunately, invertebrates are often not well represented in the fossil record, primarily due to their lack of hard structures such as skeletons from which fossils are frequently formed. Still, there was enough evidence accumulating during the 19th century to indicate that this was a significant group of organisms. The majority of the work during this century focused on classifying invertebrate fossils. Fossils

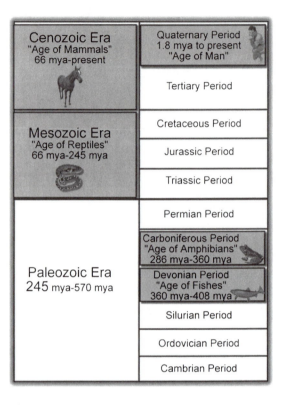

Cenozoic Era "Age of Mammals" 66 mya-present	Quaternary Period 1.8 mya to present "Age of Man"
	Tertiary Period
Mesozoic Era "Age of Reptiles" 66 mya-245 mya	Cretaceous Period
	Jurassic Period
	Triassic Period
Paleozoic Era 245 mya-570 mya	Permian Period
	Carboniferous Period "Age of Amphibians" 286 mya-360 mya
	Devonian Period "Age of Fishes" 360 mya-408 mya
	Silurian Period
	Ordovician Period
	Cambrian Period

Figure 32. A graph of the geological timescale illustrating the major ages for each vertebrate group as classified by paleontologists.

were identified that were related in form to modern organisms such as clams (Mollusks), seastars (**Echinoderms**), insects (**Arthropods**) and sponges (**Porifera**). The information provided by the 19th century invertebrate paleontologists has provided a strong foundation for modern evolutionary studies of invertebrates.

Cuvier was not the only important contributor to 19th century paleontology, although his discoveries did establish paleontology as a scientific discipline. Francois Pictet (1809–1872) authored an impressive four-volume series on paleontology between 1844 and 1846 in which he made a comparison between the fossil record and living species. This work had a strong influence on Alfred Wallace's concept of natural selection (see THEORY OF NATURAL SELECTION). Organisms other than animals were being examined during the 19th century as well. Paleontological studies of plants, sometimes called paleobotany, also developed during the 19th century and today represents an important aspect of the discipline. In addition, studies of human fossil remains marked the start of a branch of science called anthropology (see also HUMAN EVOLUTION). The diversity of paleontology toward the end of the century indicates its importance to the study and understanding of the evolution of modern species. Paleontology provides a reference of evolution, the underlying force of biological diversity.

Selected Bibliography

Bowler, Peter. *Fossils and Progress*. New York: Science History Publications, 1976.

Mayr, Ernst. *The Growth of Biological Thought: Diversity, Evolution and Inheritance*. Cambridge, MA: Harvard University Press, 1982.

Ronan, Colin A. *Science: Its History and Development among the World's Cultures*. New York: Facts on File Publications, 1982.

Taton, Rene (ed.). *History of Science: Science in the Nineteenth Century*. New York: Basic Books, 1965.

Phonograph (ca. 1857–1887): The phonograph is an instrument that permanently records sound onto some medium. There is little doubt that the phonograph is an invention of the 19th century, because there are few documented attempts of inventors trying to record sound for later playback prior to the 19th century. Yet, the invention of the phonograph is actually the product of a series of other advances during the Industrial Revolution. The manipulation of sound was an area of intense research during the 19th century. The work of Michael Faraday and others early in the century had started to define the properties of electromagnetic radiation (see ELECTROMAGNETIC THEORY). In the 1830s it became possible to use the properties of electromagnetism to send a message over long distances using the telegraph (see TELEGRAPH). Almost immediately after the invention of the telegraph, scientists and inventors began to envision a means of directly transmitting sound over a distance. By the 1870s other inventors, most notably the American Alexander Graham Bell (1847–1922), had invented the telephone (see TELEPHONE). The invention of the telephone had a strong influence on the work of one of the century's most famous American inventors, Thomas Alva Edison. He was frequently in personal communication, and often competition, with Bell. Edison and made a tremendous number of contributions to 19th century science and technology (see ELECTRIC LIGHT, ELECTRIC POWER). He is also recognized as the inventor of the first device to record and play back sound.

Edison's phonograph was not the first instrument capable of recording sound. Earlier in the century (ca. 1857) a French inventor named Leon Scott had designed a device called a phonoautograph that had the ability to record sound for analysis. This instrument lacked the ability to play back the recorded sound and was intended more as a scientific instrument. Other inventors had designed ways of representing sounds on punch cards so as to program organs and other musical instruments. Edison's invention began with his work to design a method of permanently recording the Morse code being sent over the telegraph lines (see TELEGRAPH). To do this he designed a cylinder covered with waxed paper. The incoming signal from the telegraph was transferred mechanically to a blunt needle, which made small depressions in the surface of the wax paper as the cylinder turned. The idea was that when the process was reversed and played back through a telephone speaker the operator would be able to hear the original Morse code. Unfortunately, the device was a failure. The waxed paper did not allow a great enough distinction between the dots and dashes used by operators of the Morse code. During the trials, however, Edison did notice that as the cylinder was rotated a distinct humming sound was emitted from the speaker.

By this time Edison owned a company in Menlo Park, New Jersey, that specialized in developing Edison's inventions. In 1877 Edison's group was able to design the first phonograph for the playback of sound. Rather than using waxed paper to record the sound, Edison and his team had opted for tin foil. This was not an electrical device, although Edison's lab was well involved in studies involving electricity, but rather used a hand crank for both recording the sound and playing it back. As the cylinder was turned a needle made small impressions in the tin foil. On playback the needle would vibrate from these indentions, reproducing the sound. The sound quality was not the best, but it was enough to get Edison a patent. Over the next several years he made a number of improvements to the machine, each of which served to improve sound quality. It is interesting to note that Edison did not initially have an interest in using the phonograph for music, but rather saw it either as a time-saving aide for office work such as dictation or for phonographic books for the blind. However, by the end of the century one of Edison's endeavors, the Edison Phonograph Company, was manufacturing prerecorded wax cylinders that contained 2-minute recordings of marches and other music for the general public. Edison's machines were all based on cylinders (see Figure 33), but in 1887 the modern form of the phonograph was introduced by the German-American inventor Emile Berliner (1851–1929). Berliner's phonograph, also called a gramophone, used a flat disk, often made of hardened rubber.

The phonograph remained a popular mechanism of recording music for decades following its invention. The medium on which the sounds were recorded changed from wax and rubber to vinyl and plastic, but its basic prin-

Registration of a cornet solo by the phonograph [1889]

Figure 33. An example of a 19th century phonograph. This is an early form of the device as indicated by the cylindrical shape of the recording surface. (Library of Congress photo collection)

ciples of operation remained the same. Improvements were made by the use of microphones and amplification systems throughout the 20th century. By the mid-20th century alternative forms of recording sound were being developed. Although the first form of magnetic tape was introduced at the end of the 19th century, it did not start to become a force in the recording industry until decades later. In the modern world sound may also be recorded and stored digitally on compact disks (CDs) or digital tapes. Although the technology for these recording devices is different from that of the first phonographs, the basic principle of storing sound in a permanent form for later playback remains fundamentally the same. The sound-recording industry is constantly advancing its technology to reproduce better sound using smaller packages. In many ways the industry still echoes the initial desires of Thomas Edison in the 19th century.

Selected Bibliography

Karwatka, Dennis. *Technology's Past: America's Industrial Revolution and the People who Delivered the Goods*. Ann Arbor, MI: Prakken Publications, 1996.

Read, Oliver, and Walter L. Welch. *From Tin Foil to Stereo: Evolution of the Phonograph*. Indianapolis, IN: Howard W. Sams, 1976.

Van Dulken, Stephen. *Inventing the 19th Century: 100 Inventions That Shaped the Victorian Age from Aspirin to the Zeppelin*. New York: New York University Press, 2001.

Welch, Walter L., and Leah B. S. Burt. *From Tinfoil to Stereo: The Acoustic Years of the Recording Industry 1877–1929*. Gainesville, FL: University Press of Florida, 1994.

Photography (ca. 1822–1889): The modern photographic process is a direct result of discoveries made in the early 19th century, although the concept of a camera dates from centuries earlier, possibly even to the time of the ancient Greeks. The Arab scholar Ibn-Al Haitham (965–1039), usually referred to as Alhazen, made reference in the 11th century to an early form of photographic technique that is called a *camera obscura*. The simple device consisted of a small hole in a wall. When a light was passed through the hole, a dim, inverted image could be detected and traced. These early instruments were exceptionally useful in observing eclipses and in fact are still in use by today's amateur astronomers. However, as was the case with much of Alhazen's work on optics, his discoveries were not known in the Western world and thus had to be rediscovered several centuries later. By the 16th century the camera obscura was well known to Western scientists, and by the 17th century a number of portable devices had been constructed. Throughout this time an interest developed in inventing a method of permanently recording the images.

The permanent recording of an image required that the light from the image interact with a compound to produce a chemical reaction that could distinguish between light and dark areas. This is called a photochemical reaction and is the basis of modern photography. The first scientific evidence of a compound interacting in this manner with light is the 17th century mention by the Italian Angela Sala (1576–1637) that silver nitrate turns black when exposed to sunlight. It unclear from these early experiments whether the reaction was due to the heat of the light or the light directly. This was solved in the 18th century by Johann Heinrich Shultz (1687–1744), who experimentally demonstrated that heat failed to produce the same darkening effect on silver nitrate. However, it was not until the 19th century that the process of permanently recording an image became practical.

The prime problem with the use of silver nitrate was how to preserve the image after the chemical was exposed to light. There were two aspects to this problem. The first problem was that the image produced in the silver nitrate was the reverse image of the original. In other words, the light areas of the original image were represented by a darkening of the silver nitrate and vice versa. This pattern of reversal is called a *negative*, whereas what was wanted was a *positive* of the image. The second problem, and one of greater urgency, was how to protect the unreacted silver nitrate from subsequent light exposure. Without a mechanism for protecting the light areas of the photograph the image had to be kept in the dark to protect it from degrading. In the early 19th century (ca. 1822) Joseph Nicéphore Niépce (1765–1833) invented an early form of a camera that was successful in producing an image using a silver nitrate paste applied to glass plates. However, Niépce was unable to find a mechanism of chemically altering, or fixing, the sodium nitrate so that it did not react to light after the image was recorded. To remedy this problem and produce a permanent image, Niépce designed an interim procedure by which he first produced a silver nitrate negative, and then chemically transferred the image to second piece of material as a permanent record. This was called a heliograph, and while the addition of the second step was successful in producing a positive of the image, the entire procedure frequently took considerable time.

Niépce's partner, Louis Jacques Daguerre (1789–1851) solved the light-sensitivity problem of the silver nitrate. Prior to Daguerre's work a number of people had experimented with fixing the unexposed silver nitrate using chemicals such as mercury. However, in 1837, on the advice of the German chemist John Herschel (1792–1871), Daguerre experimented with the use of sodium thiosulfate (a salt) as a fixing agent. This process greatly reduced the processing time, to less than 20 minutes, and effectively spelled the end of the heliograph as a photographic process. These early photographs were called daguerreotypes, after their inventor, and for a considerable time remained the primary method of recording images. However, the daguerreotype was not without its drawbacks, primary among of which was that the image had to be sealed between two pieces of glass to protect the silver from oxidation. One solution was to conduct the procedure using paper, and within a few short years of the invention of the daguerreotype a number of investigators were working on the problem. The most successful of these was the English inventor William Henry Fox Talbot (1800–1877). Talbot's procedure involved coating paper with silver nitrate and potassium iodide to produce a light-sensitive film. After exposure the image was developed using an acid solution and then fixed. Like the daguerreotype, the mixture had to be kept moist until it was exposed, which meant that the coating usually had to be mixed on-site. The paper process wasn't Talbot's only contribution to photography (see Figure 34). He is also recognized for producing the first book that contained photographs.

Figure 34. An early 19th century box camera such as those used by William Henry Fox Talbot. (Hulton Archive/ Getty Images)

The next major advance in 19th century photography occurred in 1878 when the American George Eastman (1854–1932) used a dry mixture mounted to celluloid. The use of dry plates predated Eastman's invention, but the invention of celluloid by the British chemist Alexander Parks (1813–1890) in 1861 enabled Eastman to mount his dry film to a lightweight backing. In 1888 Eastman invented the Kodak camera, which was followed by the 1889 invention of celluloid film. The Kodak camera, with its removable film, made photography accessible to the general public, because the film could be sent to a processing laboratory for developing. Within a few short years the Eastman Company dominated the world market in the production of photographic supplies and equipment.

The invention of this photochemical process quickly found application among both the scientific community and the general public. In both the American Civil War (1861–1865) and the Crimean War (ca. 1855) the camera was used to record the devastation of the battlefields, thus beginning the use of the camera to record important historical events. In the entertainment industry the invention of photography led to the development of the motion picture by Thomas Edison in 1889 and the subsequent birth of the cinematography industry. Photography also played an important role in a number of 19th century scientific discoveries, especially in the field of astronomy. The production of photographic plates of the night sky gave astronomers the ability to not only permanently record their observations, but it also allowed for more detailed analysis of the night sky because the astronomers did not need to complete their studies strictly during the evening hours. The American astronomer John Draper's (1811–1882) photograph of the Moon in 1839 was the first use of photography in astronomical studies. Because of photography 19th century astronomers were able to develop an early classification system for stars (see ASTRONOMY) and detect the faint motion of asteroids (see ASTEROIDS). Modern professional astronomers routinely rely on photographic plates to record their work, with direct observations of the night sky frequently performed only by amateur astronomers.

Modern photography follows much the same principles as those established in the 19th century. However, there have been a number of improvements over time, mostly in the sensitivity of the film and the performance of the associated cameras. Camera design has closely paralleled the improvements in film design. In the early 20th century instant cameras and color cameras first appeared. Throughout the century advances in camera design have allowed the specialization of cameras for almost any environment, from deep space to the depths of the ocean. In the late 20th century the invention of the digital camera initiated the next revolution in photography by allowing the instantaneous recording of images and ease of editing. However, the majority of cameras in use today still follow the photographic techniques first developed by 19th century inventors.

Selected Bibliography

Asimov, Isaac. *Asimov's New Guide to Science*. New York: Basic Books, 1984.
Gernsheim, Helmut. *A concise history of photography*. London, England: Thames and Hudson Publishers, 1965.
Gernsheim, Helmut. *The origins of photography*. London, England: Thames and Hudson Publishers, 1982.

Planetary Astronomy (1801–1898): The 17th century work of Galileo, Johannes Kepler, and Isaac Newton had firmly refuted the concept of an Earth-centered universe and in effect began the study of planetary astronomy. Yet after almost a century of intense discovery, which included the identification of satellites and calculations of planetary orbits, little attention was directed in the next century toward a further understanding of the nature of the solar system. For almost 150 years after the invention of the telescope only six planets had been identified (Mercury, Mars, Earth, Venus, Jupiter and Saturn), and many predicted that this was the extent of the solar system. For this and other reasons the majority of 18th century astronomy focused on the nature of stellar objects such as nebulae and stars. However, in the 19th century there was a renewed interest in our solar system. Innovative methods of calculating stellar distances (see ASTRONOMY) made it possible to calculate the motion of the Earth around the Sun (see EARTH), and with them came the final scientific verification that Earth was a planet. The majority of the planetary astronomy being conducted in the 19th century focused on using new approaches and improvements in astronomical methods toward identifying additional planets and satellites in the solar system.

The beginning of 19th century planetary astronomy originated with the discovery of Uranus by the English astronomer William Herschel in 1781. Although Uranus is faintly visible in the night sky, it is dim enough to be almost undistinguishable from distant stars. Using an improved telescope, Herschel was able to determine that the object moved slightly against the backdrop of stars and thus must be located within our solar system. He had discovered the seventh planet of the solar system, which attracted considerable attention over the next several decades. Furthermore, a review of historical observations indicated that data was available from as far back as 1690 and that by using this information it was possible to construct a preliminary estimate of the orbit of Uranus. It was quickly determined that its orbit was elliptical, but that when reviewed over a 130-year interval there were irregularities in the rate of movement of the planet in its orbit. The French astronomer Alexis Bouvard (1767–1843) noticed that between the discovery of the planet in 1781 and 1822 the movement of Uranus was accelerated in comparison to the historical observations. For a brief time around 1822 Uranus once again was moving at a rate consistent with expectations, but then displayed a slower rate of movement between 1822 and 1843. Since none of the inner planets displayed this phenomenon it was widely believed that something unique was happening with Uranus. Some believed that the irregularities were due to a recent collision between Uranus and a comet or large asteroid, while others suggested that the universal law of gravitation, established by Isaac Newton in the late 17th century, simply did not apply to celestial objects outside the orbit of Saturn. However, gravitation was not the problem with the orbit of Uranus, it was the key to the solution of the problem.

By the 1830s astronomers had discounted the majority of these ideas. The only plausible explanation that remained was that there was a large object, possibly another planet, outside the orbit of Uranus that was influencing the planet's rate of movement. A hint that this was possible was provided by Halley's comet. In the 17th century Edmund Halley had predicted the return of a specific comet every 76 years. However, there were minor variations in the period of the comet's orbit, which Halley attributed to gravitational interference by the outer planets, specifically Jupiter. In 1835 Halley's comet was once again delayed, but this time astronomers questioned whether a larger body outside of the orbit of Uranus was responsible. Two astronomers, Joseph Le Verrier (1811–1877) and John Couch Adams (1819–1892), independently worked out the details of the location of this unknown planet. With this information the German astronomer Gottfried Galle (1812–1910) was able to discover the planet Neptune in 1846.

Having found one planet by examining irregularities in a known planet's orbit, astronomers next turned their attention to the inner solar system. For some time astronomers had recognized that the orbit of Mercury, the innermost planet, was irregular. As Mercury approached its perihelion, or closest approach to the Sun, the planet accelerates slightly due to the gravitational pull of the other planets. However, the magnitude of this pull can't be explained completely by Newtonian gravitational theory. In 1846 Le Verrier proposed that another planet was located in a closer orbit to the Sun, and that the mass of this planet was responsible for Mercury's observed behavior. Le Verrier named this planet Vulcan and immediately a number of astronomers began the search for the planet. Although a number of observers claimed to have detected Vulcan, a positive identification of the suspected planet eluded astronomers for more than 70 years. In 1916, the American physicist Albert Einstein provided an explanation for Mercury's motion as part of his general theory of relativity. In effect, Einstein explained that the intense gravitational pull of the Sun was actually warping the fabric of space in the vicinity of Mercury, causing it to have varied rates of acceleration. The search for the planet Vulcan represents the transition from traditional Newtonian physics to Einstein's more modern ideas. Thus, although unsuccessful, the search for the planet represents an important achievement in the study of physics and astronomy.

The study of planetary astronomy was not limited solely to identifying new planets. The invention of photography had revolutionized the study of astronomy (see ASTRONOMY, PHOTOGRAPHY), and in the 19th century a group of astronomers were turning their attention toward detecting

Table 5
Satellites Discovered during the 19th Century

Planet	Satellite	Year	Discoverer
Mars	Deimos	1877	Asaph Hall
	Phobos	1877	Asaph Hall
Jupiter	Almalthea	1892	Edward Barnard
Saturn	Hyperion	1848	William Bond
			William Lassell
	Phoebe	1898	William Pickering
Uranus	Ariel	1851	William Lassell
	Umbriel	1851	William Lassell
Neptune	Triton	1846	William Lassell

additional satellites around the planets. A list of these 19th century discoveries is provided in Table 5.

Planetary astronomy of the 19th century also finally provided an accepted explanation on the nature of Saturn's rings. First detected by Galileo in the 17th century, the rings were first thought of to be solid spheres surrounding the planet. Later improvements in telescope design suggested that the rings were not solid, but rather consisted of a series of smaller rings in discrete orbits around the planet. However, this provided a real challenge to Newton's concept of universal gravitation, because over time the rings should coalesce into small moons. The French astronomer Simon-Pierre Laplace (1749–1827) suggested that since the rings varied in their distance from the planet they were subjected to different levels of gravitational attraction. He believed that this would suffice to inhibit the formation of larger bodies. In 1855 James Clerk Maxwell mathematically demonstrated that the rings must consist of small particles, each in a unique orbit around the planet. The mathematician Edouard Roche (1820–1883) provided further support for this idea when he mathematically demonstrated that a satellite must be a minimum distance from a planet to escape being destroyed by the gravitational tidal forces of the planet. This distance, called the Roche limit, is approximately 2.44 times the radius of the planet. Because Saturn's rings are within the Roche limit, they most likely represent the remains of a small moon that was destroyed by Saturn's gravitational field. The fact that the rings are indeed small particles was confirmed by the Pioneer and Viking space probes of the 20th century.

The study of planetary astronomy in the 19th century represents a significant milestone for the astronomical sciences. Since the invention of the telescope in the 17th century the majority of discoveries had been based on direct observation of the solar system. However, by the 19th century the use of mathematics and physics had become almost routine in astronomy. Therefore astronomers were able not only to discover a new planet, but also to test some developing theories on the physical nature of the universe. This was the beginning of the study of astrophysics, which would play an important role in 20th century science.

Selected Bibliography

Baum, Richard, and William Sheehan. *In Search of Planet Vulcan: The Ghost in Newton's Clockwork.* New York: Plenum Trade, 1997.

Hoskin, Michael. *The Concise History of Astronomy.* New York: Cambridge University Press, 1999.

Littmann, Mark. *Planets Beyond: Discovering the Outer Solar System.* New York: John Wiley & Sons, 1988.

Motz, Lloyd, and Jefferson Hane Weaver. *The Story of Astronomy.* New York: Plenum Press, 1995.

Printing Press (1811–1884): For much of recorded history civilizations had documented important events and transactions by hand. In the earliest times this involved the use of paintings, stone engravings, papyrus rolls and clay tablets. The invention of paper by the Chinese in the 2nd century provided an easier and lighter medium for recording information. The production of books, which were previously collections of scrolls or tablets, was greatly enhanced by the invention of paper, although for almost seven centuries books were copied by hand. As a result books were extremely valuable and most people of those times rarely even saw a book. Around 770 the Japanese were using a process called block printing, which uses a carved block of wood to print letters, a process that probably originated with the Chinese a century earlier. Within a hundred years the Chinese and Koreans were producing books using this simple printing style. The idea of using movable type to facilitate a more rapid printing process was first invented by the Chinese around 1045 and was widely used by the Koreans within 200 years. Over the next several centuries the Chinese and Koreans experimented with various forms of type, including those made from wood and bronze. A printing press with removable type was first introduced into Europe in the 1400s by a German inventor named Johann Gutenberg (ca. 1397–1468). In a style similar to that of the Koreans, although he probably was unaware of the

Korean press, Gutenberg used metal letters to prolong the life of the type and developed an oil-based ink to reduce smearing. Using a style of press utilized by the wine industry to process grapes, Gutenberg designed a printing press that applied even pressure to the printed page. In 1456 he produced 200 copies of the Bible, of which 47 have survived to modern times. More important, he initiated the formation of a publishing industry that would have a tremendous influence on science, education and culture over the next several centuries.

The Industrial Revolution of the late 18th and 19th centuries demanded a change in the process by which goods, including literature, were produced. The process of manually setting type into a frame and then individually printing the pages was faster than copying by hand, but it still remained a relatively slow system of producing printed material. Inventors were working on ways of reducing the processing time for printing, thus decreasing cost and increasing demand for their product. In the 19th century there were two major advances to the printing press that increased the efficiency of production: the invention of a process of setting type called linotype and the invention of high-speed cylinder presses for commercial use.

The 19th century was a time of tremendous evolution in printing press design. Over the course of the century a number of improvements were made to the Gutenberg design of printing press, most of which targeted increasing the speed of production. The most successful design used a cylinder to rotate the type into contact with the paper, rather than use a lever to press the type onto the paper, as was the case with most pre-19th century presses. A number of inventors attempted to design a workable cylinder press, but credit for the first successful machine is given to Friedrich Koenig (1774–1833). The primary problem with the design of cylinder machines before Koenig's work was in the inking mechanism. Koenig and his partners designed a series of machines that attempted to solve these problems, the first of which they patented in 1811. They settled on a vertical inking system that used a piston to continuously supply ink to the face of the print. The newspaper industry had a strong interest in Koenig's work, and in 1813 he settled on a design that was used in the production the *Times* of London in 1814. By 1816 Koenig had invented a steam-powered machine that had the ability to print on two sides of paper and could print about 1,000 sheets of material an hour. The industrialization of the printing system was well under way.

Over the next several decades a number of improvements were made to the cylinder design, and this was reflected in an increase in the efficiency of the machines. One of the more successful designs was invented by the American inventor Richard Hoe (1812–1886), who introduced a machine called a type-revolving press. Hoe belonged to a family of inventors who made a series of improvements to the cylinder design of press. One of the enhancements involved imbedding the print directly onto the cylinder, instead of simply using the cylinder to move the paper over the type. But his greatest improvement was in the overall efficiencies of the machines. The type-revolving press was widely used by the newspaper industry because it had the capability when introduced of producing more than 10,000 copies per hour. By the end of the century the Hoe machines were able to print more than 70,000 pages an hour. When coupled with the linotype system (see later), the result was a highly efficient, inexpensive means of producing printed material for the general public.

Since the invention of the printing press the setting of type by hand was the greatest obstacle to the mass production of printed material. The advances in printing press design by the 19th century had made the process of delivering the ink to the paper more efficient. Newspaper editors, book publishers and inventors all recognized that the next major reform in the printing industry would have to accelerate the speed by which the page was assembled before printing. The German inventor Ottmar Merganthaler (1854–1899) was the first to successfully design a mechanic typesetting device that could compete against human hands. Merganthaler began experimenting with prototypes of his machine in the 1870s, but did not produce the first successful working model until 1884. This machine, called the Second Band Machine, consisted of a typewriter station (see TYPEWRITER) for the operator where the individual lines of type were entered. When a key was depressed on the keyboard, a mold was lowered into place, which was subsequently filled with hot metal, producing what was commonly called a slug but which actually was an impression for a single line of type. The machine was called a linotype, because it produced single lines of type, which were then combined to form a printed page. Many historians of the communication industry consider this to be one of the most important inventions in the history of the industry in that it has a significant impact on the cost and speed of producing printed material of all types.

There can be little doubt that the improvements in the printing industry during the 19th century directly benefited both Western and eventually world culture, but also the scientific and engineering communities who rely so heavily on dissemination of information to promote their research. Whereas books and printed literature were relatively rare at the start of the century, by the start of the 20th century millions of books were being produced on a yearly basis. The development of computer systems, word processing software and digital printing processes has expanded on the abilities of the printing press to communicate to the masses.

Selected Bibliography

Karwatka, Dennis. *Technology's Past: America's Industrial Revolution and the People Who Delivered the Goods.* Ann Arbor, MI: Prakken Publications, 1996.

Karwatka, Dennis. *Technology's Past, Vol. 2: More Heroes of Invention and Innovention.* Ann Arbor, MI: Prakken Publications, 1999.

Moran, James. *Printing Presses: History and Development from the Fifteenth Century to Modern Times.* Berkeley: University of California Press, 1973.

Sterne, Harold, E. *A Catalogue of Nineteenth Century Printing Presses.* New Castle, DE: Oak Knoll Press, 2001.

R

Radio (1847–1901): The invention of radio, or the wireless tele-graph as it was initially called, represents one of the more significant discov-eries in communication in human history. As was the case with many of the technology-related discoveries of the late 19th century, the invention of radio was based on the pioneering efforts of physicists early in the century to under-stand the nature of electromagnetism and the electromagnetic spectrum. The early work on electromagnetism by Hans Oersted, Michael Faraday and André Ampère demonstrated that a distinct relationship exists between elec-trical and magnetic forces and that an electrical field could be used to induce a magnetic field and vice versa. Around 1847 the German physicist Hermann van Helmholtz suggested that the movement of electricity was due to the oscillation of the force, although he did not provide experimental proof. In 1865 James Clerk Maxwell summarized decades of work on electromagnet-ism when he developed the mathematical basis of the electromagnetic theory (see ELECTROMAGNETIC THEORY). One of the more important aspects of this theory was the idea that electricity moves through space as an oscillating wave. However, the majority of Maxwell's work was theoretical. The experimental verification of the electromagnetic theory, and the begin-nings of the invention of radio, started with the work of Heinrich Rudolf Hertz, a German physicist.

As a scientist Hertz had a strong interest in the study of Maxwell's elec-tromagnetic theory. In 1887 he designed an experiment to determine the wavelength, or the distance between the peaks (see Figure 35), of an electro-magnetic wave. First Hertz developed an electrical circuit that discharged a current through a series of wires. This had the effect of producing a constant electrical oscillation at a wavelength dependent on the configuration of the circuit. The next step was to design a means of detecting the signal coming

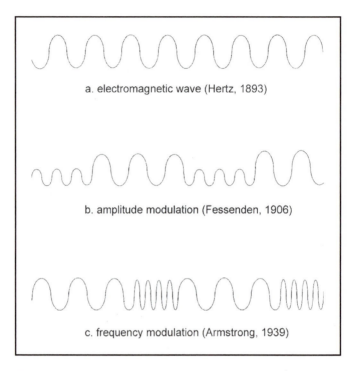

a. electromagnetic wave (Hertz, 1893)

b. amplitude modulation (Fessenden, 1906)

c. frequency modulation (Armstrong, 1939)

Figure 35. An illustration of electromagnetic waves with the 20th century discoveries of amplitude and frequency modulation.

from the generator. For this Hertz designed a detector consisting of a single piece of rectangular wire. The wire was not continuous, but rather had a small break in one side. When the detector received the signal it would produce a spark at the gap. After some experimentation with the length of the wire Hertz was able to design a detector with which he could not only detect the presence of the wave but also indicate the intensity of the signal and the duration of the spark. Once the system was developed Hertz set out to examine the properties of electromagnetic waves. By moving the detector various distances from the oscillating source and observing the intensity of the spark across the gap in the detector, Hertz was able to estimate the wavelength of the electromagnetic wave.

In a later series of experiments Hertz was able to estimate the velocity of the wave, which he determined to be the same as the speed of light. Hertz published a number of influential papers on the nature of electromagnetic waves, including a book entitled *Electric Waves* (1893). Hertz's experiments are significant not only because they provided the theoretical foundation for the invention of radio, but also because they provided an important verification of Maxwell's electromagnetic theory (see ELECTROMAGNETIC THEORY).

Hertz did not exploit the use of electromagnetic waves for communication, but their application was being explored by a number of other inventors and scientists at the end of the century. As early as 1866 the American inventor Elis Loomis received a patent from the United States government for the design of a wireless device that would use kites to raise a wire about 600 feet for the purpose of wireless communication. Unfortunately the setup was never constructed. Another proponent of wireless communication was the English scientist William Crookes, who suggested around 1892 that electromagnetic waves could be adapted for use with Morse code. Crookes was also the first to suggest that different wavelengths could be utilized to penetrate buildings and clouds. It soon was recognized that to communicate over larger distances a larger and improved detector was needed. A number of variations of Hertz's detectors were explored, but in 1895 Oliver Joseph Lodge (1851–1940) invented the first radio antenna. This allowed him to detect radio waves more than a mile from the source.

The first patent that resulted in the design of a successful wireless radio was awarded in 1897 to the Italian inventor Guglielmo Marconi. Using the principles proposed by Hertz, Marconi constructed a wireless radio to transmit Morse code a distance of about 1.5 miles. Within the next few years Marconi had designed a wireless communication network that connected England and France and in 1901 succeeded in broadcasting a simple signal across the Atlantic Ocean. This demonstration effectively commenced the modern age of radio communication. It is interesting to note that although the term *wireless communication* is now recognized as indicating signals traveling through air, the initial idea was to transmit an electric signal through bodies of water such as rivers. This idea was fostered by the early problems with underwater telegraph cables (see TELEGRAPH), but some suggested that it could be used more extensively for communication. The procedure was not widely accepted at the time, but modern navies use this form of communication to send messages to submarines. Using extremely low-frequency (ELF) signals, short messages can be transmitted to underwater submarines, thus decreasing the need for these vessels to risk detection on the surface. This became especially useful during the Cold War with the invention of nuclear-powered submarines that had the capability to remain submerged for long periods of time.

Although modern radios are governed by the principles of the electromagnetic theory (see ELECTROMAGNETIC THEORY), there have been a number of significant improvements since Marconi's first signal was transmitted. In 1906 the Canadian scientist Reginald Fessenden (1866–1932) recognized that rather than use a series of dots and dashes to send messages, it was possible to use a continuous signal whose amplitude varied in proportion to the sound that it was carrying (see Figure 35). This marked the beginning of amplitude-modulated radio, commonly called AM radio. While this marked the beginning of true radio transmissions, it was possible for atmospheric phenomena to produce interference, or static, in the AM signal. In 1939 the American physicist Edwin Armstrong (1890–1954) developed a method of modulating the frequency of a radio wave to send a message. In frequency-modulated, or FM, sound is imposed on a high-frequency carrier wave by increasing or decreasing the wavelength of the carrier to correspond to the audio signal (see Figure 35). Although limited to line-of-sight (LOS) communications only, FM radio represents the primary means of delivering radio signals around the globe. Radio signals may now be relayed using satellites, thus increasing the distance that they can effectively be transmitted. It should also be noted that radio waves may be emitted from natural sources. Astronomers now frequently study the radio emissions of distant stars and other interstellar objects.

The invention of the radio in the 19th century had a strong influence on the development of mass communication in the 20th century. Although originally overshadowed by the telegraph, the radio became a dominant form of communication by World War II and would remain the primary method of communicating electronically until the widespread development of the television in the mid-20th century.

Selected Bibliography

Archer, Gleason, L. *History of Radio to 1926.* New York: Arno Press, 1971.

Bordeau, Sanford P. *Volts to Hertz. . . . The rise of Electricity.* Minneapolis, MN: Burgess Publishing, 1982.

Karwatka, Dennis. *Technology's Past, Vol. 2: More Heroes of Invention and Innovation.* Ann Arbor, MI: Prakken Publications, 1999.

Winston, Brian. *Media Technology and Society, A History: From the Telegraph to the Internet.* London, England: Routledge, 1998.

Rubber (1823–ca. 1880): The modern industrialized world uses rubber for a wide variety of purposes, from motor vehicle tires to insulation. Although rubber has been around for centuries in South America, it was not until the 19th century that it was modified into its present form. Rubber is a chemical derived from the South American plant *Hevea brasiliensis*, a tree widely found in the Amazon region of Brazil. One of the fluids from this tree, commonly called latex, was used by the ancient Mayan civilization as early as 1600 B.C.E. Latex is found in a wide variety of plant species but is almost unusable in its natural form. However the Mayans developed a method of heating the latex from *H. brasiliensis* and mixing it with other compounds to create a manageable solid that was shaped into rubber balls and figures. The term *rubber* is a product of the 18th century, when the chemist Joseph Priestly discovered that it was possible to use the compound to erase, or rub out, pencil marks. Within a few years it was also discovered that rubber, when mixed with turpentine, produced a waterproof substance that could be used as a protective coating for clothing. After a series of improvements in the early 19th century (1823) the Scottish chemist Charles Macintosh patented a rubber-treated cloth that was used as rain gear.

The use of natural rubber was not without problems. Rubber does not retain its elastic properties well in extreme temperatures. At cold temperatures natural rubber becomes brittle and difficult to bend, whereas at high temperatures it begins to lose its solid nature and becomes sticky and gumlike. In the early 19th century industrial chemistry, or the making of products for commercial sale, was gaining popularity (see also NITROGLYCERINE). Some chemists believed that they could modify rubber to make it more stable at the high and low ends of the temperature scales. One of these was the American industrialist Charles Goodyear (1800–1860). Although not a chemist by training, Goodyear was convinced of the usefulness of rubber and made a series of attempts to enhance its stability. One of his early experiments (1837) was to add nitric acid to latex. The result was a more stable product, but it still had limitations. However, in 1839 Goodyear stumbled on a better solution. While working in his kitchen, which doubled as his laboratory, Goodyear dropped a piece of rubber treated with sulfur onto a hot stove. To his amazement when he removed the rubber from the stove it had not turned sticky as was the case previously with heated rubber. When the heat-treated sulfur-rubber combination was brought into the cold it did not turn brittle. Goodyear had discovered vulcanized rubber. Over the next several years he experimented to find the ideal sulfur-rubber ratio and temperature of heat treating and in 1844 received a patent for his discovery. Despite the importance of his discovery, and the well-recognized company that holds his name, vulcanized rubber was not an immediate success and Goodyear earned little profit from his discovery.

Rubber trees are common in the Brazilian rainforest, but they are difficult to harvest owing to the inhospitable terrain. In the 1870s plantations of *H. brasiliensis* trees were established on Sri Lanka and in southeast Asia by British, French and Dutch companies. *H. brasiliensis* grows well in tropical climates and was easily transferred to these areas. Demand for rubber increased significantly after the invention of the automobile in the late 19th century. By the 1920s the majority of rubber was being harvested from plantations and not the natural rubber forests.

Rubber is an interesting chemical compound whose function (in the form of latex) in the plant is not completely understood but is believed by some to help repair damage to the plant caused by predators. The investigation of the chemical properties of rubber were initiated in the 19th century. In 1826, Michael Faraday, an influential 19th century English chemist and inventor, discovered that latex is a hydrocarbon (see CHEMISTRY). Later in the century chemists determined that rubber is composed of numerous isoprene subunits. Isoprene (C_5H_8) is an interesting molecule in that when isoprene units are linked together there is elasticity in their chemical bonds, giving rubber its unique characteristics.

Because production of rubber is limited to areas that are within 700 miles of the equator, there has historically been some concern among industrialized nations on the reliability of crucial rubber supplies during wartime. While it is possible to recycle rubber, as was discovered toward the end of the 19th century, the automobile industry requires a significant amount of raw material annually. In the early years of World War II the German government experimented with developing synthetic rubber from petroleum products, but with little success. However, since petroleum is a hydrocarbon it should be possible to convert it to isoprene to make rubber. After the capture of the rubber plantations by the Japanese in World War II the American government began an aggressive campaign to develop synthetic rubber, and by 1942 the Americans were manufacturing a significant portion of their rubber needs. More recently, with the increase in oil prices and an uncertainty over world supply of both oil and rubber, there has developed an interest in developing nonpetroleum sources of synthetic rubbers. Botanists are also looking for other sources of natural latex that may be used as a substitute to traditional sources. There have been some reports that the guayule plant of Mexico may be an alternate source for rubber. Regardless of its future, the discovery of vulcanized rubber marks an important advance for industrial chemistry during the 19th century.

Selected Bibliography

Asimov, Isaac. *Asimov's New Guide to Science*. New York: Basic Books, 1984.

Karwatka, Dennis. *Technology's Past: America's Industrial Revolution and the People Who Delivered the Goods*. Ann Arbor, MI: Prakken Publications, 1996.

Van Dulken, Stephen. *Inventing the 19th Century: 100 Inventions That Shaped the Victorian Age from Aspirin to the Zeppelin*. New York: New York University Press, 2001.

Wilson, Charles Morrow. *Trees & Test Tubes: The Story of Rubber*. New York: Henry Holt, 1943.

S

Seismograph (1839–1889): A seismograph is an instrument designed to measure the movement of the Earth's crust during geological events such as earthquakes. While earthquakes themselves have been known since ancient times, the origins of which were frequently a source of myth, it was not until the 19th century that an effort was made to scientifically quantify the intensity of an earthquake for study. However, there were earlier attempts to detect earthquakes, especially in areas of the Far East where earthquakes are common. In the 2nd century the Chinese mathematician Heng Chang (78–142) is reported to have invented a primitive form of a seismograph that consisted of a series of dragon heads, each containing a ball. Many people in this region at this time believed earthquakes to be caused by subterranean fights between dragons. When an earthquake occurred a pendulum would knock the ball loose from the dragon's head. Although this instrument lacked any ability to quantify earthquake intensity, it did provide a general indication of the direction of the crust's movement. Western science began to develop an interest in earthquakes following a strong quake that struck Lisbon in 1755. In 1760 the English geologist John Michell (1724–1793) proposed that earthquakes were due to the movement of heated water far beneath the surface and that these disturbances traveled as waves through the earth's crust. While he had little scientific evidence for his ideas, he did foster an interest in studying earthquakes. His ideas are considered by many to have initiated the science of seismology, a discipline of the geological sciences that remains very active in the modern scientific world.

The science of seismology began to take shape in the 19th century with the invention of an instrument called the seismograph. Since the 17th century a number of scientists had designed primitive instruments to measure earthquakes based on the motion of a pendulum. The refinement of these devices continued into the 19th century, with improvements being continuously made in the sensitivity of the seismometers as well as their ability to permanently record the observations. One of the first of the 19th century seismometers was invented by the Scottish geologist James Forbes (1809–1868). This device had the ability to permanently record the movements of the pendulum on a piece of paper, although the sensitivity of the instrument was very limited. In 1855 the Italian physicist Luigi Palmieri (1807–1896) developed a seismometer that used a mercury-filled tube to detect movements in the earth's surface. The movement of the mercury within the tube completed an electronic circuit, which in turn activated an electromagnetic switch that stopped a clock, thus recording the time of the motion. This instrument represented a significant increase in sensitivity compared to earlier, pendulum-based seismometers, but it could not distinguish between vibrations associated with road traffic from actual earthquakes.

In 1880 the English geologist John Milne (1850–1913) invented the first seismometer that resembles those in use today. Milne's seismometer was the result of his studies of earthquakes in and around Japan. To eliminate the influence of local vibrations, Milne anchored his seismograph directly into the bedrock. Attached to the support was a weak spring, from which two large blocks were suspended. One of these blocks was oriented to detect motions of the earth's crust in a north-south direction, and the other was oriented east-west. When the bedrock vibrated, the movement was detected by the spring and in turn was recorded by a pen. By the end of the century Milne had designed a worldwide network of 13 recording seismographs. Milne suggested that this network should be able to detect an earthquake anywhere in the world and thus is credited for the design of the first seismological network. In 1889 his theory was proved when seismometers in Germany detected an earthquake that had occurred in Tokyo a little more than an hour earlier.

In the early 20th century the major improvements to the seismometers were in sensitivity. Work was focused primarily on reducing the friction between the recording pen and paper, because friction reduced the ability of the instrument to detect small motions in the earth's crust. Various methods were tried, including using smoked paper. Modern seismometers frequently use light beams that create images on specially formatted paper, although pen recorders are still found in many places. In 1935 the American scientist Charles Richter (1900–1985) designed a system of quantifying the intensity of earthquakes. The Richter scale uses a series of numbers between 0 and 9 Each increase of one number represents a 10-fold increase in the movement of the seismometer, which corresponds to roughly a 30-fold increase in the released energy. This scale is the one used to report earthquake intensity internationally.

Seismometers still play an important role in modern science. There are currently several hundred seismometers worldwide dedicated to the detection of earthquakes, but the instruments have found other applications as well. Geologists and seismologists frequently use specialized seismometers to search for underground deposits of fossil fuels such as oil. Seismometers are also used internationally by defense agencies to detect underground tests of nuclear weapons. Although the form of each of these instruments varies slightly, the basic function and structure are the same as those of the first seismometers invented in the 19th century.

Selected Bibliography

Asimov, Isaac. *Asimov's New Guide to Science*. New York: Basic Books, 1984.
Krebs, Robert E. *Scientific Laws, Principles and Theories: A Reference Guide*. Westport, CT: Greenwood Press, 2001.
Lapedes, Daniel N. (ed.). *McGraw-Hill Encyclopedia of the Geological Sciences*. New York: McGraw-Hill Book, 1978.
Oldroyd, David R. *Sciences of the Earth: Studies in the History of Mineralogy and Geology*. Algershot, UK: Ashgate Publishing, 1998.

Spectroscope (1803–1900): During the early 19th century physicists were beginning to define various forms of electromagnetic radiation. In the process they started to explore the spectrum of electromagnetic energy (see ELECTROMAGNETIC SPECTRUM). Investigations of the visible spectrum of light had begun during the Scientific Revolution of the 17th century, when Isaac Newton used a prism to demonstrate that color is a property of light and not of the object being illuminated. In the early 19th century there was a renewed interest in the study of light, although most of the research during this time was directed toward the determination of whether light existed as a wave or a particle (see LIGHT). The study of the wavelengths of visible light played an important role in these investigations, and the primary instrument of these early studies remained the prism. However, as the century progressed physicists recognized that they required a more specialized instrument for the study of the light spectrum. The result was the invention of an instrument called the spectroscope, which quickly found application in a diverse array of scientific investigations.

The beginnings of spectral analysis may be said to have started with the work of Joseph von Fraunhofer. Fraunhofer had an interest in optics, and specifically in studying the refraction of different wavelengths of light. In one of his experiments (1814) Fraunhofer used a prism to break apart sunlight. In doing so he noticed that the spectrum of wavelengths was not continuous, but consisted of a series of distinct bands separated by dark lines or regions. While this was not entirely a new discovery, since scientists earlier in the century, most notably William Hyde Wollaston, had recorded similar observations, Fraunhofer's work was significant in that he was the first to map these lines. This was possible as a result of his manufacturing of high-quality prisms that were relatively free of impurities. Using one of these prisms, Fraunhofer was able to perform a detailed analysis of the banding pattern of sunlight. After determining that the spectral lines were not an artifact of his prism or experimental design, Fraunhofer was able to identify about 570 distinct bands in the solar spectrum. The spectral lines of the sun are commonly called Fraunhofer lines in his honor.

Although the scientific community was initially unable to explain the origin of the spectral lines, they had previously used the phenomena to develop theories on the physical nature of light. For example, around 1803 Thomas Young began a series of experiments using an early form of spectral analysis. His observations, when coupled with the work of later physicists, eventually demonstrated that light had the properties of a wave (see LIGHT). The lack of a scientific explanation for the Fraunhofer lines did not deter chemists from accumulating a significant amount of data on the spectrum of substances other than sunlight. In 1845 William Allen Miller (1817–1870), an English chemist, published the first diagrams of spectra and by 1855 scientists such as the Frenchman Jean Foucault were suggesting that elements such as sodium may be present in the sun. In the decades following Fraunhofer's death scientists began to recognize that the spectra of the sun was probably due to the types of gases it contained.

In the mid-1800s the German chemist Robert Wilhelm Bunsen began to use spectral analysis for the study of chemistry. Bunsen's research suggested that it was the elements within a compound that, when burned, produced the characteristic spectral pattern. Bunsen was the first to make detailed studies of spectra, but the recognition that the spectral pattern is dependent on the chemical composition of a flame did not truly originate with him. Around 1826 the English scientist William Henry Fox Talbot had suggested that chemicals such as platinum, potassium and sulfur gave flames a distinct color and even suggested that flame (spectral) analysis may be used for the study of the chemical properties of compounds. Unfortunately, during Talbot's time science lacked the ability to experimentally perform such experiments. However, to facilitate his research Bunsen invented a high-intensity burner that allowed small amounts of an element to be burned. Using this apparatus, commonly called a Bunsen burner, Bunsen was able to identify two new elements, rubidium and cesium, and in the process began a new era of discovery for chemical elements (see ELEMENTS). The Bunsen burner is now a common instrument of any chemical laboratory and has changed little since the mid-19th century.

To further improve spectral research Bunsen and the German physicist Gustav Robert Kirchhoff invented an instrument called a spectroscope to facilitate their research in spectral analysis. Although early forms of spectroscopes had been in existence since the early part of the century, their invention (ca. 1860) represents the basic design of the more modern form of the instrument. The structure of the Bunsen-Kirchhoff spectroscope was relatively simple. First a tube focused the light from a source to a prism. The light was divided into its principle wavelengths, or colors, as it passed through the prism. The light from the prism then passed through a second tube to the eye of the observer, thus isolating the spectrum from external light interference. Using Bunsen's burner it was possible to generate relatively pure spectrums for the elements. Thus, when Bunsen and Kirchhoff trained their spectroscope on the sun, they were able to detect the presence of distinct elements. While their findings that the Sun had many of the same elements as the Earth represented an important discovery, of greater importance was the beginning of a field of science called astrophysics (see ASTRONOMY), a discipline that in the 20th century would revolutionize the understanding of the chemical nature of the universe. Kirchhoff used the spectroscope to develop some tentative explanations on the origins of the Fraunhofer lines, but a more precise explanation would have to wait until 1870. In that year James Clerk Maxwell published *Theory of Heat*, which included not only a description of the molecular nature of gases (see KINETIC THEORY OF GASES) but also an explanation that the spectral lines were the result of molecular motion. Others would refine this idea, but physicists were on the way to recognizing the relationship between spectra and molecular motion.

It should be noted that spectral analysis was not confined solely to the spectrum of electromagnetic radiation being emitted by a heated object. As part of his kinetic theory of gases, Kirchhoff recognized that the dark lines of the spectrum were the result of the wavelengths of light that were being absorbed by an object. He suggested that some objects, called black bodies, may absorb all wavelengths of light but that when they are heated may emit all wavelengths. Although Kirchhoff's theory on black body radiation did not play an important role for 19th century physics, it represented an exceptionally important turning point for the study of physics. In 1900 the German scientist Max Planck (1858–1947) provided the mathematical proof of Kirchhoff's ideas. This proof, recognized as the beginning of the quantum theory of physics, included the **Planck's constant** for energy and is considered to be the start of modern physics. In fact, the presence of matter that absorbs all wavelengths of light, frequently called dark matter, resurfaced in late 20th century astrophysics in an attempt to explain the structure of the universe.

The 19th century spectroscope still finds application in many areas of modern science. Many introductory science laboratories utilize a spectroscope to analyze the visible spectrum of light. However, most scientific investigations of light now require a quantification of the results. An instrument called a mass spectrophotometer operates on much the same principles as a spectroscope, but has the added ability to measure the precise position of spectral lines. Other versions of spectrophotometers have the ability to analyze a single wavelength of light. The spectrophotometer has found a wide variety of scientific applications, from the study of the composition of astronomical objects such as stars, to determining the precise molecular components of chemical compounds.

Selected Bibliography

Bud, Robert, and Deborah Jean Warner (eds.). *Instruments of Science*. New York: Garland Publishing, 1998.

McGucken, William. *Nineteenth-Century Spectroscopy*. Baltimore, MD: The Johns Hopkins Press, 1969.

Park, David. *The Fire within the Eye: A Historical Essay on the Nature and Meaning of Light*. Princeton: Princeton University Press, 1997.

Steam Engines (1803–1897): The power of steam to drive mechanical devices has been recognized for more than 2,000 years. The ancient Greeks constructed an instrument called an aeoliphile, which consisted of a hollow sphere with several bent tubes projecting from its surface. Steam was delivered to the inside of the sphere from a cauldron, and as it escaped through the bent tubes the entire device began to spin. However, the first steam instrument that was actually used for scientific purposes was a water-commanding device invented around 1663 by the Englishman Edward Somerset. For some time prior to his work many had predicted that it could be possible to move water by using the power of steam. Somerset's device used steam to create a vacuum, which in turn was utilized to pump water. This engine had the capability of raising a column of water 40 feet. The English inventor Thomas Savery (ca. 1650–1715) is credited with the design and patent for the first modern steam engine in 1698. The applicability of steam engines was quickly realized by industry and over the next two centuries considerable improvements were made to their power and reliability. In 1712 an English engineer named Thomas Newcomen (1663–1729) invented the first piston-powered steam engine. Later in the century James Watt, a

Scottish engineer, made a number of improvements to the steam engine and in the process revolutionized the use of steam power, which in turn effectively began the Industrial Revolution. In the 19th century the steam engine would once again benefit both science and industry, as well as play an important role in the scientific investigation of nature.

Scientific investigations frequently produce technological breakthroughs that benefit society or industry. However, the study of steam engines in the 19th century had the reverse effect. Although steam engines had been in widespread use since Newcomen's time, the scientific basis of how they worked, specifically in the transfer of heat, was poorly understood. In the 1820s, the French engineer Nicolas Sadi Carnot began a series of detailed investigations involving the operation of Watt's steam engine. Watt's design utilized two chambers at different temperatures to amplify the generation of steam power and is frequently called a condensing engine. While a significant improvement over the Newcomen engine, it was not clear as to how the condensing engine generated work. At the start of the 19th century most of the scientific community considered heat to be a fluid substance called caloric (see HEAT) and the movement of heat between two objects was due to the transfer of caloric. This also applied to the operation of the steam engine, with the power of the engine believed to be in a direct relationship to the abundance of caloric within the steam. In fact, many physicists had accepted Watt's law, which stated that for a given amount of steam there was an equal amount of caloric at all temperatures. In fact, many compared the operation of a steam engine to that of a water wheel, the amount of water drives the wheel. Carnot's studies suggested that this was not the case, although he did not dismiss the role of caloric.

Carnot was one of the first to propose that it was not the amount of steam that was important in the operation of the engine, but rather the heat of the fire that generated the steam. As the temperature of the burner was increased, the was a subsequent increase in steam pressure. This corresponded to an increase in the potential for work. Although still a supporter of the caloric theory, which was now considered a mechanism by which heat was moved, Carnot correctly identified the relationship among the temperature of the burner, the pressure of the steam, and the amount of work produced. These ideas were presented in Carnot's 1824 publication, *On the Motive Power of Fire*. In this work he also suggested that a perfect steam engine would have the ability to transfer heat from its origination point to the site of the work without loss. Furthermore, he suggested that the amount of energy in the universe was constant and was merely being converted between forms. Carnot's thoughts remained unrecognized for almost a decade after their publication, but these ideas played an important role in the development of the laws of thermodynamics later in the 19th century (see LAWS OF THER-MODYNAMICS). In addition, his work helped improve the efficiency of the steam engine. Carnot had demonstrated that the transfer of the heat was the driving force behind the operation of the steam engine, so it was possible to increase efficiency by increasing the temperature difference between the condensers of Watt's engine. An example of a 19th century steam engine is shown in Figure 36.

Steam engines are recognized as an important advance for the 19th century and not only because of their contributions to the study of physics. A series of improvements in steam engine technology also resulted in the development of new forms of powered transportation. At the end of the 18th century steam engines had been adapted for maritime use with the invention of the steamboat in 1783. By the early 19th century a number of steamboats were in operation, including the American industrialist Robert Fulton's (1765–1815) successful enterprise the *Clermont*. Soon thereafter, some inventors began turning their attention to the development of steam-powered railways. There is evidence of railway use in Western industry from the time of the 16th century. However, the majority of these examples were horse powered and provided limited service to the mining industry. The first successful steam

Figure 36. An example of an early 19th century steam engine. (Library of Congress photo collection)

locomotive was designed by the Englishman Richard Trevithick (1771–1833) in 1803. However, it was not until 1814 that the English inventor George Stephenson (1781–1848) demonstrated the applicability of steam engines to rail service. Steam engines originally were designed for use by heavy industry, typically mining, but by the mid-19th century their usefulness for hauling a wide variety of cargo had been realized. By the end of the century the prime method of moving both people and goods was by rail, and the steam locomotive would dominate the transportation industry until the invention of the automobile and airplane (see INTERNAL COMBUSTION ENGINE).

However, there was one more advance in steam technology before the start of the 20th century. In 1884 the English inventor Charles Parsons developed what is considered to be the first successful version of a modern steam turbine. This first steam turbine differed from the existing steam engine in that as the steam is used to drive a series of movable blades, which in turn are attached to a rotor. There were several earlier examples of steam turbines in the 19th century, most notably the machines developed for use in a number of industries, including saw mills. However, these machines were very unstable and did not find widespread application. Parsons's machine extracted the pressure of the steam using a series of stages, which greatly increased the efficiency of the turbine. In 1897 Parsons's was successful in equipping a ship, the *Turbinia*, with a steam turbine. This was the first large-scale maritime application for the steam turbine, and several the world's marine powers, especially the Royal Navy, were quick to recognize the importance of the steam turbine.

By the end of the 19th century steam engines had been transformed from simple steam generators to instruments capable of producing enough work to power a 40,000-ton ship. In the process of its development, the steam engine became an important factor in the industrialization of Western cultures because of its ability to produce large amounts of energy. In addition, the steam engine was instrumental in developing an understanding on the nature of heat and thermodynamics. It was not until later in the 20th century that the superiority of the steam engine would be challenged by the invention and development of the internal combustion engine. However, recently there has been a renewed interest in steam engine technology because it can generate power without producing high levels of air pollution.

Selected Bibliography

Hills, Richard L. *Power from Steam: A History of the Stationary Steam Engine.* Cambridge, UK: Cambridge University Press, 1989.

Karwatka, Dennis. *Technology's Past, Vol. 2: More Heroes of Invention and Innovation.* Ann Arbor, MI: Prakken Publications, 1999.

Robbins, Michael. *The Railway Age*. Manchester, UK: Mandolin Press, 1998.
Spangenburg, Ray, and Diane K. Moser. *The History of Science in the Nineteenth Century.* New York: Facts on File, 1994.

Steel (1854–1875): The Iron Age began in the area of the Caucasus Mountains around 1500 B.C.E. Within a few centuries the technology for removing metallic iron from iron ore and manufacturing weapons and tools had spread throughout much of the Middle East. The process involves placing iron ore in a furnace with a preheated carbon source. Initially this carbon source was charcoal, which is wood that has been heated to a point where it is primarily carbon with very little oxygen. When the charcoal is heated along with the iron ore the carbon in the charcoal uses the oxygen in the ore for combustion. The result is iron. Initially the furnaces in which this process was conducted were not hot enough to melt iron (which occurs at about 2,800°F or 1,538°C). The result of the lack of sufficient heat was a softened iron called wrought iron that could be bent or hammered into useful shapes. By the 14th century it was possible to heat iron ore to its melting point, and it could then be poured into special molds to make cast iron. This enabled a much higher level of flexibility in the design of iron products, but cast iron was a brittle substance that could not withstand high amounts of stress. Another problem was that it was difficult to reach the required temperatures using charcoal and by the start of the 18th century wood was becoming a rare resource in certain parts of Europe. A useful solution to this problem was presented in 1709 when the Englishman Abraham Darby (1678–1717) recognized that coal could also be used as a fuel, provided that the oxygen was first removed from it in the same manner as for charcoal formation. The result was a compound called coke, which continues to play an important role in iron and steel production.

Steel is a high-grade iron that contains less than 1% carbon versus the approximate 4% carbon content of cast iron. When the carbon content is within this range, the steel has a strength much greater than that of wrought or cast iron. Unfortunately, the coke being used to heat the iron ore to the desired temperature frequently added impurities, often sulphur and phosphorous, that ruined the formation of the steel. Before the 19th century it was possible to make steel using traditional methods, but it was almost a hit-or-miss process and thus the amount of steel produced was typically very small. For a considerable period steel was considered to be a rare metal, an indication of how difficult the manufacturing process was.

In 1854 the English inventor Henry Bessemer (1813–1898) began to work on a solution to the problem of how to produce large quantities of steel. His interest in the issue began after he discussed a new type of military projectile with the leader of France, Napoleon III. This projectile had special rifling carved into it that would enable it to travel a greater distance to the target with a higher degree of accuracy. Unfortunately, it also needed a larger charge to propel it, and Napoleon questioned whether the existing cast iron cannons would be able to withstand the launching of the projectile. To make his projectile marketable, Bessemer had to first develop a more reliable metal for the casing of the gun and the logical choice was steel. It is interesting to note that before taking on this challenge Bessemer had little knowledge of the steel-making process. Bessemer's solution was to supply an extra burst of oxygen to the molten cast iron (also called pig iron) to burn off the carbon and other impurities to the desired level. Most of the experts at the time thought that the adding of air to the molten metal would cool the metal rather than remove excess carbon, but Bessemer demonstrated that the addition of the oxygen-rich air increased the level of combustion, elevating the temperature to the range necessary to reduce the carbon content to the desired levels. In 1856 Bessemer announced the invention of the blast furnace, which employed this process.

The problem of impurities, namely phosphorous, continued to plague the manufacture of steel, even when it was formed in a blast furnace. One solution, and the one used by Bessemer, was to import iron ore with a low phosphorous content from other regions of the world, namely Sweden. But the importing process was not inexpensive and an alternative method was needed. Several inventors proposed mechanisms of reducing the phosphorous content. One of these was to conduct the entire process in an open hearth in which a combination of iron ore and pig iron was heated. A more effective method was to design the furnace so that during the process of the reaction the phosphorous was "pulled" from the steel. This was first effectively demonstrated in 1875 when Sidney Thomas lined the interior of the furnace with a combination of manganese and limestone. This combination removed most of the phosphorous from the iron ore. It was now possible to produce large quantities of high-quality steel from almost any available source of iron ore.

It is also possible to specialize steel for specific functions. These specialized compounds are commonly called steel alloys. A large number of steel alloys are on the market, each of which adds a specific metal to the steel-making process to give the steel added protection against heat, pressure or rust. Stainless steel is an example of a steel alloy. First invented by Harry Brearley in 1913, this steel alloy contains nickel and chromium, which serve to protect the iron from the oxidative damage of the atmosphere. Stainless steel is now commonly found in kitchen appliances and implements. However, despite the versatility of steel, it remains a heavy metal to transport. Some other metals are lighter than steel, but there had traditionally been problems isolating the metal from its metallic ores. One example is aluminum, which can be separated from its ore using an electrical current, but the amount of energy required made it a prohibitive process. In 1886 the American Charles Hall (1863–1914) and the French Paul Louis Héroult (1863–1914) independently discovered a process by which aluminum ore was first mixed with another chemical called cryolite, which reduced the amount of energy required for purifying the ore. Although this process did not have much of an influence on 19th century society, aluminum and similar metals have played a strong role in the manufacture of a wide variety of items in modern society.

The invention of an effective steel manufacturing process in the 19th century marks a major advance for the science of metallurgy, but it also represents a major technological advance that had a tremendous influence on society. The availability of inexpensive steel empowered the Industrial Revolution of the late 19th and early 20th centuries. A number of technological enhancements have been made in the production of steel since the 19th century, but the basic process first worked out be Bessemer remains the same, a tribute to the ingenuity of the engineers and scientists of the time.

Selected Bibliography

Asimov, Isaac. *Asimov's New Guide to Science*. New York: Basic Books, 1984.
Fisher, Douglas Alan. *The Epic of Steel*. New York: Harper & Row, 1963.
Hornsby, Jeremy. *The Story of Inventions*. New York: Crescent Books, 1977.

The Sun (1814–1908): As the central star of our solar system the Sun has played an important historical role in the study of astronomy. In the early 17th century early astronomers began to accumulate evidence that our solar system was not centered on the Earth, but rather was a Sun-centered system. In many ways this ideas marked the beginnings of the Scientific Revolution, which focused on conclusions based on experimentation rather than on philosophical discussions. This change in effect provided the foundations for 19th century scientific thinking. In addition, 17th century astronomers made some early calculations as to the mass of the Sun and its distance from the Earth, and even began to suggest that the composition of the stars may actually be similar to that of our Sun. However, the true beginnings of solar astronomy started with the work of the English astronomer William Herschel in the 18th century. Herschel designed a new generation of high-powered telescopes, which enabled him to discover the planet Uranus in 1781 (see PLANETARY ASTRONOMY). He not only scientifically explored the concept that the stars and the Sun were the same, but also demonstrated that the Sun itself moved through space, effectively ending the idea that Earth is the center of the universe. His work, coupled with a new age of astronomers, continued into the 19th century as the Sun became a crucial testing ground for understanding the nature of the solar system.

Since the time of the ancient Greeks, astronomers had considered the Sun to be a perfect sphere and thus a direct manifestation of the creator. The invention of the telescope in the early 17th century dramatically changed this perception. In some of the earliest uses of the telescopes (ca. 1610) astronomers noticed small blemishes in the surface of the Sun, although credit for the first true detection of sunspots belongs to Chinese astronomers who had first recorded them centuries earlier. The 17th century Italian astronomer Galileo presented evidence that these were not planets moving across the face of the Sun, as many thought, but rather spots on the surface of the Sun that moved according to the Sun's rotation. However, despite this discovery, little attention was given to the study of sunspots until the 1820s when the German astronomer Heinrich Schwabe (1789–1875) started examining them as possible candidates for the planet Vulcan, believed to exist inside of the orbit of Mercury (see PLANETARY ASTRONOMY). Although he was unsuccessful in locating the nonexistent planet, his two-decade search indicated a distinct pattern to sunspot activity. In 1843 Schwabe recorded that the activity peaked about every ten years, with a modern value established to be closer to eleven years. The relationship between sunspot activity and terrestrial magnetism was made several years later after Alexander von Hemboldt (1769–1859), a German natural historian, suggested an importance to the cycle. Around 1852 John Lamont (1805–1879) and Edward Sabine (1788–1883) independently noted that there was a direct correlation between sunspot activity and the intensity of the Earth's magnetic field. In 1908 the American scientist George Ellery Hale (1868–1938) made the link when he detected strong magnetic fields being emitted from sunspots. It should be noted that the timing of the sunspot cycle is not exact. As early as 1893 astronomers had noticed that historically there had been long periods, as much as seven decades, of little of no sunspot activity. However, explanations as to why this occurred would have to wait for significant advances in solar astronomy in the late 20th century.

Although the Sun was imperfectly shaped, most of the early 19th century scientific community still considered it to be eternal and virtually indestructible. However, the discovery of the properties of heat and the laws of thermodynamics in the mid-19th century drastically changed this perception (see HEAT, LAWS OF THERMODYNAMICS). Heat now abided by specific physical laws, and as the ultimate source of heat the Sun must be bound by the same principles. It was obvious that the Sun was capable of converting energy between forms, as suggested by the first law of thermodynamics. However, according to the second law of thermodynamics this conversion was inefficient, which meant that the Sun must be constantly losing its heat-generating ability. The ultimate source of this energy, nuclear fusion, was unknown until the mid-20th century. Some 19th century astronomers, such as Hermann von Helmholtz and William Thomson, suggested that the gravitational forces of the Sun were causing it to constrict in size, which resulted in the output of energy detected on Earth. In 1854 Helmholtz used this principle to calculate the age of the Sun to be 25 million years and suggested that it would exhaust its fuel within another 10 million years. This theory presented a temporary solution for astronomers and physicists, but it created a far greater problem for geologists and biologists, who required a significantly greater age of the Earth for their developing theories (see EARTH, GEOLOGICAL TIME, GEOLOGY, THEORY OF ACQUIRED CHARACTERISTICS, THEORY OF NATURAL SELECTION).

The invention of several new instruments in the 19th century, specifically the spectroscope and the photographic plate, resulted in a revolution in the study of astronomy (see PHOTOGRAPHY, SPECTROSCOPE). Since the time of Isaac Newton it had been recognized that sunlight consisted of distinct wavelengths of light that represented color. By using a prism it was possible to break light into its distinct wavelengths. A significant advance in spectral analysis began around 1814 with the work of the German scientist Joseph von Fraunhofer. Fraunhofer developed a high-quality prism for analyzing the spectrum. When he used this prism to break sunlight into its principle wavelengths, he discovered that the spectrum of colors was not continuous. Rather, it contained distinct bands, separated by dark lines (see SPECTROSCOPE). Spectral analysis of elements by Robert Bunsen and Gustav Kirchhoff (ca. 1859) indicated that each element had a distinct spectrum (see ELEMENTS). When this information was applied to the spectral analysis of the Sun physicists quickly discovered that the Sun was composed more than fifty known elements. Using these principles the Swedish physicist Anders Angström (1814–1874) determined in 1862 that the Sun consisted primarily of hydrogen, the most abundant element in the universe. In 1891 George Hale invented a device called a spectroheliograph, which enabled him to photograph specific wavelengths of the solar spectrum and discover the presence of calcium in the Sun's atmosphere. Furthermore, the discovery of spectral analysis meant that it was now possible to determine the precise chemical composition of stars, regardless of their distance from Earth. These events marked the beginning of an important phase in astrophysics as scientists used the Sun as a testing ground for the development of tools to analyze the stars.

By the end of the 19th century solar physicists had another tool at their disposal, the photographic plate (see PHOTOGRAPHY). Using photographic plates it was possible to freeze individual moments of time for detailed analysis at a later date. In the mid-1800s solar physicists had a renewed interest in observing solar activity such as sunspots and solar eclipses. Eclipses enabled scientists to study the composition of the outer layers of the Sun, called the corona, without the interference of light. These analyses indicated that the Sun was a violent place, with solar flares, sunspots and intense activity. As the century drew to a close a number of observatories were dedicated to the study of the Sun.

The study of solar astrophysics continued well into the 20th century. In fact some aspects of the Sun's activity remain unexplained by modern physicists. However, the solar discoveries of the 19th century provided important information on the nature of the Sun. Prior to the 19th century the Sun was frequently considered to be perfect in both composition and function. The discoveries during this century indicated not only that the Sun comprised the same elements as found elsewhere in the universe, but also that it is a dynamic object whose phases directly influence life on Earth. A large portion of the modern work in astrophysics focuses on Sun-Earth interaction. Sunspots and times of increased solar output have been known to adversely effect telecommunications on Earth. NASA currently has a number of spacecraft that monitor the activity of the Sun, including the Hubble space telescope. They are also actively looking for sunlike stars in other areas of our galaxy to develop a better understanding of our position in the universe. Our understanding of the Sun originates with the work of these 19th century physicists and astronomers.

Selected Bibliography

Asimov, Isaac. *Asimov's New Guide to Science.* New York: Basic Books, 1984.

Haufbauer, Karl. *Exploring the Sun: Solar Science Since Galileo.* Baltimore: Johns Hopkins University Press,1991.

Hoskin, Michael. *The Concise History of Astronomy.* New York: Cambridge University Press, 1999.

Motz, Lloyd, and Jefferson Hane Weaver. *The Story of Astronomy.* New York: Plenum Press, 1995.

Synthetic Dyes (ca. 1820–1875): Dyes serve a wide variety of purposes in the industrialized world. Since ancient times the textile industry has used natural dyes, obtained from extracts of plant or animal tissue, to produce cloth of various colors. The use of dyes was also of interest to biologists since the invention of the microscope in the 17th century. Microscopes function by passing light from a source through a specimen and then onto one or more magnifying lenses. For this process to be effective the specimen must be very thin. Early microscopes lacked the ability to produce high levels of magnification, but by the 18th century there was enough of an improvement in the design of the microscope so that early microbiologists could begin to distinguish the internal structure of cells. Unfortunately, in most cases cells lack a significant amount of color, which greatly hindered the ability of the researchers to resolve internal structures. Prior to the 19th century a number of natural dyes were used by scientists, including indigo and blueberry juice. By the early 19th century several German researchers, namely Christian Ehrenberg (1795–1876) and Joseph von Gerlach (1820–1896) had demonstrated that dyes could be used to differentially stain the internal components of cells. Gerlach manufactured several dyes from natural substances and is regarded by many as a pioneer in the field of histology. However, in the early 19th century scientists generally lacked effective dyes and techniques to make the process widely acceptable.

The first synthetic dye was manufactured accidentally by William Henry Perkin (1838–1907), an English chemist. Perkin was working in the laboratory of the German chemist August von Hofmann (1818–1892), who had an interest in manufacturing the antimalaria drug **quinine**. Hofmann's lab was using coal extract as a source for the organic material needed in the synthesis of quinine. Hofmann's laboratory had previously been successful in isolating a group of compounds called anilines, and there was some hope that anilines could be chemically modified to produce the profitable drug. Quinine had been isolated from the bark of a South African tree in the 1820s by the French chemists Pierre Pelletier and Joseph Caventou (see CELL BIOLOGY), who were also successful in determining the atomic structure of the compound.

In the mid-1850s the study of organic chemistry was in its infancy and the structure of many organic molecules, including quinine, had yet to be determined. When Perkin examined the atomic composition of a chemical called ally toluidine, he concluded that it had approximately the same atomic composition as quinine. It was missing only a few oxygen molecules, which Perkin thought that he could add by using the chemical potassium dichromate as an oxygen source. The result was a complete disaster. Not only had Perkin failed in producing quinine, but the reddish-brown material left over was worthless. In a second attempt he combined the chemicals aniline and potassium dichromate and got basically the same result. However, when he attempted to wash the material with alcohol, a brilliant purple color was produced. This was aniline purple, the first of the synthetic dyes. Perkin applied for a patent for his invention in 1856.

Aniline purple was almost immediately commercially successful in the textile industry and resulted in a number of chemists pursuing industrial applications for their work in organic chemistry (see ORGANIC CHEM-ISTRY). Perkin continued his success in synthesizing organic compounds and in 1875 synthetically produced a compound called coumarin, the first synthetic perfume.

Within a few years of the production of aniline purple a number of additional dyes had been produced, as well as the means of chemically producing natural dyes in the laboratory. Not only did these dyes benefit the textile industry, but also cell biologists who were making tremendous advances in the identification of the internal organelles of plant and animal cells (see CELL BIOLOGY). This tradition continues to this day. Modern cell biologists have available a wide variety of artificial dyes that can differentiate the internal structures of a cell. These dyes are now known to interact with specific portions of the cell, or even individual biomolecules within the cell membranes (see BIOMOLECULES). A special class of synthetic compounds is the florescent dyes, which enable cell biologists to examine the internal structures of cells, such as the cytoskeleton, using special microscopes. The modern advances in cell biology, specifically as they relate to the pharmaceutical industry, are a direct result of the first of the 19th century industrial chemists who investigated the process by which compounds such as dyes could be made artificially.

Selected Bibliography

Asimov, Isaac. *Asimov's New Guide to Science*. New York: Basic Books, 1984.

Magner, Lois N. *A History of the Life Sciences*, 2nd ed. New York: Marcel Decker, 1994.

Van Dulken, Stephen. *Inventing the 19th Century: 100 Inventions That Shaped the Victorian Age from Aspirin to the Zeppelin*. New York: New York University Press, 2001.

T

Telegraph (1837–1866): Long before the 19th century, military communications required the ability to signal important messages over long distances. Signal fires were the earliest form of long-distance communications and their use predates the time of the ancient Greeks. Centuries later, signals were transmitted over long distances using flashes of light reflected from mirrors. These instruments, called heliographs, were later perfected to transmit signals over significant distances. Although heliographs were useful, they required sunlight and it was difficult to send anything but primitive messages. At the close of the 18th century an improved signaling mechanism relied on the positioning of crossed poles to indicate various letters of the alphabet. These poles, called semaphores, were placed on towers and could be viewed over much longer distances and in a variety of weather conditions using telescopes. The study of static electricity in the late 18th century, and the recognition that it could be transmitted over a metal wire, suggested that an electric signaling mechanism might be able to transmit messages. By the end of the century a number of devices had been constructed that used single wires for each letter of the alphabet. Messages could be sent over longer distances, but the technology did not exist to insulate the wires correctly, creating a significant amount of interference between the wires. As a result most of these systems were abandoned.

The 19th century invention of the single-wire electric telegraph was made possible by earlier advances in the study of electromagnetic theories and principles. The most significant of these was the recognition not only that the forces of electricity and magnetism were related, but also that magnetic fields could generate an electric potential and vice versa (see ELECTRO-MAGNETIC THEORY). One of the instruments designed to study electricity was the galvanometer. Early galvanometers were simple devices that indicated the presence of an electric current by the deflection of a needle. Within a few years of this discovery (ca. 1820) a few inventors were experimenting with the possibility of using a galvanometer to detect a signal being sent over a wire. Some of these early instruments used multiple wires, as was the case with the static electricity signaling devices mentioned earlier, while others used variations in the deflection of the needle to indicate the letters of the alphabet.

By the early 1830s a number of experimenters were independently working on single-wire electric telegraph systems. The simultaneous nature of their work is a strong indication that the time was right for the development of the telegraph. However, first a few technical obstacles had to be overcome. An electrical signal degrades over distance, and this presented a severe problem for the design of a long-distance instrument. This problem was partially solved in 1837 by the English scientist Edward Davy (1806–1885) with his invention of an "electrical renewer." The electrical renewer used a battery to enhance the level of the signal and in doing so greatly increased the efficiency of electrical communications. Another important advance was the invention of the electric relay by the American physicist Joseph Henry in 1837. This relay consisted of an armature attached to a magnet that was tightly wound with fine wire. When a signal arrived over an incoming wire to the magnet the armature would be pulled against the magnet. This connection transferred power to a recorder, usually some sort of printing device. Henry realized that he had all of the components to construct a working telegraph, but he failed to pursue the opportunity. Henry went on to become the first director of the Smithsonian Institution.

Credit for the invention of the modern form of the telegraph is frequently given to Samuel Morse (1791–1872), an American artist by training. Morse's telegraph functioned by the opening and closing of an electrical connection at one end of the circuit. At the other end of the circuit was a pen that was activated by the current in the wire. The initial configuration of the pen put the pen in contact with the paper at all times. Incoming signals caused the pen to deflect, producing a record that looked like one from modern heart-monitoring equipment. After several modifications Morse constructed a pen that came in contact with the paper only when a signal was received. The result was a series of dots and dashes on the paper. In 1838 Morse developed a code that assigned a unique combination of dashes and/or dots for each letter of the alphabet. This is called the Morse code and represents a simple, but effective mechanism of sending messages over long distances. The idea of a standardized code had existed since the earliest experiments with the telegraph. Using Henry's ideas for improving signal strength, Morse was soon

Figure 37. An illustration of the telegraph system used between Baltimore and Washington in 1844. (Library of Congress photo collection)

successful in developing a telegraph that could send signals over significant distances (see Figure 37). The usefulness of the telegraph was demonstrated during the 1844 American presidential elections, when Morse was successful is sending the results of the vice presidential nomination to Washington hours before it was officially received by rail. The American government quickly constructed a telegraph line between Washington and Baltimore, which was in operation by the end of 1844. The modern age of long-distance communication had begun.

It is important to note that at about the same time other inventors in other countries were working on similar devices. For example, in the 1830s the English inventor Charles Wheatstone (1802–1875) invented a telegraph system that was capable of sending a signal over a distance of about a half mile. In 1837, before Morse's work, he received a patent from the government of England for his invention. Similar instruments were being developed in other areas of Europe. However, it was Morse's version, and his early success in demonstrating the usefulness of the telegraph in communication, that was most widely accepted.

One of the biggest challenges to the deployment of a large-scale telegraph network was the design of underwater cables. Early attempts to place telegraph cables underwater quickly demonstrated that more effective insulation mechanisms were needed, because the constant exposure to water resulted in an electrical failure within the line. Various new materials for insulators were attempted, including glass, wax and various resins, with little success. In the 1840s a latex compound called gutta-percha was isolated from a Malaysian tree. Gutta-percha is an incredibly water-resistant substance and provided the necessary insulation material. In 1849 the first underwater cable was stretched across the Connecticut River and was quickly followed by cables across the English Channel and the Hudson River. After these successes investors turned their eyes to a transatlantic telegraph cable. After several failed attempts in 1858, the American businessman Cyrus Field (1819–1892) was able to complete the cable in 1866. Rapid, transcontinental communication between Europe and America was now possible.

There can be no doubt that the exact origins of the telegraph are clouded in history, with a large number of inventors in various countries developing similar local versions. However, as is the case with all inventions that have multiple origins, the invention of the telegraph is a demonstration of the important link between basic scientific research and technological advance. As was the case with many discoveries of the late 19th century, a series of scientific studies, in this case specifically the investigations into the properties of electromagnetism and electricity, enabled a technological advance for society, which in turn changed the culture of the society. Through the work of Davy, Henry, Morse and Wheatstone, long-distance communication, although still in its early stages, was possible by the end of the 19th century. Thus, the invention of the telegraph represents the starting point of this technology for our modern electronic society.

Selected Bibliography

Archer, Gleason, L. *History of Radio to 1926.* New York: Arno Press, 1971.
Bordeau, Sanford P. *Volts to Hertz . . . The rise of Electricity.* Minneapolis, MN: Burgess Publishing, 1982.
Winston, Brian. *Media Technology and Society, A History: From the Telegraph to the Internet.* London, England: Routledge, 1998.

Telephone (ca. 1876): The invention of the telephone in the later decades of the 19th century was the result of almost 75 years of scientific research. Since the early 1800s physicists had been investigating the properties of electromagnetic waves and formed many of the modern theories on electromagnetic radiation (see ELECTROMAGNETIC THEORY). For example, in the early years of the century physicists has established the connection between electrical and magnetic forces and recognized that these forces are interchangeable. Even before James Clerk Maxwell's establishment of the electromagnetic theory in 1875 (see ELECTROMAGNETIC THEORY) and its groundbreaking definition of the physics of electromagnetic forces, inventors and scientists had recognized that it was possible to transmit electrical messages over a wire. Entrepreneurs and inventors quickly realized the importance of this discovery and by the 1830s the telegraph had been invented, which marked the beginning of the long-distance communication industry (see TELEGRAPH). Within just a few decades a series of similar scientific advances in the understanding of the wavelike nature of electromagnetic radiation led to the invention of the wireless telegraph, or radio (see RADIO). Between the invention of the telegraph and wireless radio communication, scientists and inventors were actively investigating how to use wires to transmit sound over long distances.

The invention of the first telephone to successfully send sound over a wire is frequently credit to the American inventor Alexander Graham Bell, although there is evidence that a number of other inventors, most specifically the successful American telegraph owner Elisha Gray (1835–1901), were developing similar instruments at about the same time. Bell was interested in improving the telegraph so that multiple messages could be transmitted over a single wire. To do this he was experimenting with the concept of sending Morse code messages using different tones, which could then be deciphered at the receiving end. This idea was called a harmonic telegraph, a prototype of which had previously been constructed by Gray. These sounds were to be generated mechanically by metal reeds that vibrated at a specific tone. The idea was to pass the vibration itself down the wire, but when Bell and his assistant Thomas Watson (1854–1934) began their experiment they discovered that the actual sound was being transmitted.

Once it became obvious that sound could be directly transmitted, Bell and Watson became interested in the invention of a device that would allow simultaneous two-way communication. To function, the instrument needed to convert sound waves to a fluctuating electrical current, which could then be reconstructed into sound waves at the receiving end. Bell experimented with several different methods of making this conversion, but eventually settled on the idea of using a small tube connected to a cup of dilute sulfuric acid. When the sound moved through the tube it struck the acid. The acid was connected to the terminal of a battery to provide the electrical current. During one set of experiments Bell spilled some of the acid on himself, and called out to Watson, "Mr. Watson, come here! I want you." Watson, who was in an adjoining room at the time, heard Bell clearly over their new invention. This marked the invention of the telephone, for which Bell received a patent from the United States government in 1876.

The importance of the telephone for long-distance communication was quickly recognized by investors, and within a year the Bell Telephone Company had been established to provide telephone service. However, the sulfuric acid telephone originally designed by Bell was short-lived. Within a year Thomas Edison, an American inventor responsible for the widespread distribution of electric power (see ELECTRIC POWER), had made a substantial improvement to the telephone. Edison substituted a carbon powder for the sulfuric acid. The powder was easier to contain and conducted a clearer electric signal when compressed by an incoming sound, such as a voice. Edison's carbon transmitter was quickly adopted by Western Union, a telegraph company that was becoming a strong competitor for the newly created telephone industry. But Bell Telephone countered with Watson's invention of a magnetic transmitter. This competition had the dual result of not only promoting relatively rapid improvements in the invention but also in the marketing of telephones to the common citizen, increasing both the popularity of the invention and its impact on the public.

There is no doubt about the importance of the telephone on society. By the end of the 19th century companies providing telephone service crossed the entire continent and Europe. The telephone provided a relatively inexpensive means of long-distance communication. Despite the fact that it was invented over a century ago, the telephone remains the most popular means of conducting personal business over a distance. However, the technological importance of Bell's invention would not have been possible without the scientific foundation in the fields of electricity and electromagnetism. As the 19th century drew to a close a new trend was emerging in the scientific community. As was the case since the beginning of the Scientific Revolution, the scientific community remained responsible for establishing scientific breakthroughs. However, more frequently inventors were developing new consumer products without a full understanding of the scientific basis of their inventions. For almost a century the basic design of the telephone remained unchanged, with the majority of changes being implemented in increasing signal clarity and speed of connection. However, in the late 20th century improvements in wireless technology and the increased availability of satellite communication systems resulted in another revolution in the telephone. Although most homes still have a telephone system that utilizes the basic technology introduced in the 19th century, an increasing number of consumers are utilizing wireless systems to remain connected to business and families.

Selected Bibliography

Archer, Gleason, L. *History of Radio to 1926.* New York: Arno Press, 1971.
Grosvenor, Edwin S., and Morgan Wesson. *Alexander Graham Bell: The Life and Times of the Man Who Invented the Telephone.* New York: Harry N. Abrams,1997.
Winston, Brian. *Media Technology and Society, A History: From the Telegraph to the Internet.* London, England: Routledge, 1998.

Theory of Acquired Characteristics (ca. 1800–ca. 1859): The process by which organic organisms evolve in response to environmental pressures is the core theme in the study of biological science. The concept of evolution had been in existence for a considerable amount of time prior to the 19th century, and there is evidence that philosophers as far back as the ancient Greeks had recognized that it could be possible for organisms to change over time. After the start of the Scientific Revolution in the 1600s the study of geological processes and fossils each suggested that the surface of the planet was dynamic (see GEOLOGY, PALEONTOLOGY), and thus many naturalists thought it logical to suspect that living creatures must have to respond to these environmental changes or become extinct. However, it was not until the 19th century that naturalists began to develop scientific theories of how evolution occurred, specifically regarding the mechanism by which species changed over time. Opposing these new concepts was a group of scientists and influential religious organizations that contested that each species had remained unchanged since their creation. This later viewpoint had held considerable influence for centuries as to how many naturalists viewed the place of a species in nature. This was primarily the result of the lack of a unifying theory from the scientific community and because many scientists were trained and supported by religious schools. This began to change in the early years of the 19th century, when the French naturalist Jean Baptiste Lamarck proposed a mechanism for evolution. Lamarck's development of the theory of acquired characteristics was significant not necessarily for the validity of its ideas, since most of the premises of his theory were widely disproved by Charles Darwin later in the century (see THEORY OF NATURAL SELECTION), but rather for opening the door for the scientific investigation of evolution and for the subsequent influence it had on evolutionary thought for the remainder of the 19th century.

Before Lamarck developed his theories on evolution he had a reputable career studying and classifying invertebrate organisms. In this capacity he was provided with the opportunity to observe a wide diversity of invertebrate organisms, specifically mollusks, from museums and fossil collections. Lamarck noticed that not only were there examples of gradual change in the invertebrates he studied over time, but there were also fossils for which he could find no closely related living organisms. Of course, the expanse of geologic time had yet to be established in the early years of the 19th century (see GEOLOGICAL TIME), but Lamarck was convinced that the process of evolutionary change was gradual and thus not immediately noticeable. However, the problem of the extinct species initially troubled Lamarck, as it did many naturalists of his time. Many possible reasons for extinction were being debated in the early 19th century, including the action of a biblical flood or the possibility that some remnant of the species may still be alive and undiscovered in an isolated area of the planet. None of these ideas were very popular with Lamarck. From his studies of the invertebrate fossil record Lamarck began to think that what appeared to be an extinct species from the fossil record may simply have changed over time to a more modern form. Lamarck recognized that the geological conditions of the planet varied over time and realized that organisms would also have to vary their form to adapt to these changes.

It was this philosophy on species extinction that led Lamarck to propose his theory of acquired characteristics in his book *Zoological Philosophy* (1809). Contained within this work are Lamarck's two laws of evolutionary change. Both of these ideas had been in some format for a considerable amount of time, but it was Lamarck who brought them together and thus attracted the attention of the scientific community. The first of these is commonly called the law of use and disuse. This law states that the more that an organism uses an organ, the more the organ will strengthen and develop over time. Similarly, organs that are not in use will weaken and eventually disappear. His second law states that the changes caused by the law of use and disuse are conserved from generation to generation. This means that organisms can pass acquired characteristics on to their offspring. In Lamarck's view this allowed for relatively small changes to occur over long periods of time. This second law is commonly called the law of acquired characteristics, although it involves the principles of use and disuse. Lamarck believed that these changes were not only physical, but behavioral as well. He gave several examples of his theory, including the commonly given example by which giraffes acquired longer necks by stretching for leaves higher in the trees and then passing that trait to their offspring. However, Lamarck never experimentally tested his ideas, most likely because he thought that the process took a considerable amount of time. He also never suggested the means by which these traits were transmitted at the cellular or organismal level. There were some strong critics of Lamarck's ideas. The French scientist Georges Cuvier, a pioneer in the science of paleontology (see PALEONTOLOGY), suggested what he thought to be a simpler explanation for the fossil record. According to Cuvier, all organisms had been created together, but a series of catastrophic events eliminated species over time. In this theory of catastrophism (see PALEON-TOLOGY), species did not change over time. Cuvier possessed considerable influence in French science and his theory was considered the rival of Lamarck's for several decades.

Lamarck's recognition that the environment has a role in the process of evolution is an important advance in the study of evolutionary science. However, the prime difference between Lamarck's ideas and the theory of natural selection supported by most biologists is that Lamarck believed that the environment caused the variation in the organisms, while Darwin's theory proposed that the environment acts on, or selects, from preexisting variations in the population. Furthermore, Darwin believed that all organisms were of a common lineage, but Lamarck believed that each organism had varied since creation to the modern forms in existence today. It is interesting to note that although many considered Darwin to be strongly opposed to the theories of Lamarck, in many ways Darwin's early theories on pangenesis are compatible with the inheritance of acquired characteristics (see EMBRYOLOGY). Thus, as was the case with the remainder of evolutionary thinking in the 19th century, Charles Darwin was revising his theories over time as well. Lamarck was not alone in his ideas. Darwin's grandfather, the English naturalist Erasmus Darwin held views similar to those of Lamarck and without doubt had some influence over Charles Darwin's ideas of evolution. The Lamarckian concept of evolution effectively ended with Charles Darwin's publication of *The Origin of Species* in 1859. Over the next decade the theory of natural selection gained favor among evolutionary scientists, although not without serious debate (see THEORY OF NATURAL SELECTION), and by the end of the century most of the support for Lamarck's ideas had faded.

However, Lamarck's ideas have not completely faded into the past. In the mid-20th century the Russian biologist Trofim Denisovich Lysenko (1898–1976) resurrected Lamarck's principles to suggest that if a generation of grain seed was soaked in cold water, all subsequent generations of seed from those plants would be cold resistant. Despite the lack of experimental evidence, and the fact that his ideas were scientifically invalid, Lysenko's Lamarckian viewpoints dominated Russian agricultural practices for almost three decades. However, microbiologists in the late 20th century began to realize that there could be examples of Lamarckian evolution in nature, specifically among bacteria and possibly in the antibody genes of the human immune system. In each of these cases it may be possible for inherited characteristics to be passed on to the next generation, although a significant amount of research must still be performed in this area. Lamarck's work is frequently overshadowed by that of Charles Darwin, but the theory of acquired characteristics represents a significant milestone for the study of biology and evolution in the 19th century.

Selected Bibliography

Bowler, Peter. *Evolution: The History of an Idea*. Berkeley: University of California Press,1989.
Burkhardt, Richard W., Jr. *The Spirit of System: Lamarck and Evolutionary Biology*. Cambridge, MA: Harvard University Press,1995.
Krebs, Robert E. *Scientific Laws, Principles and Theories: A Reference Guide*. Westport, CT: Greenwood Press, 2001.
Mayr, Ernst. *The Growth of Biological Thought: Diversity, Evolution and Inheritance*. Cambridge, MA: Harvard University Press, 1982.
Steele, Edward J., Robyn A. Lindley, and Robert V. Blanden. *Lamarck's Signature: How Retrogenes Are Changing Darwin's Natural Selection Paradigm*. Reading, MA: Perseus Books, 1998.

Theory of Natural Selection (1809–1871): The process of evolution represents the changes in a species over time. Modern science recognizes that species change in response to competition and the environment, but for the majority of the history of science many considered a species to be unchanging since the time of creation. For example, the ancient Greek philosopher and naturalist Aristotle was a proponent of the Chain of Beings, which ranked life in varying orders of complexity. Although Aristotle used the Chain of Beings as a classification system, until the 18th century many viewed this ladder of life, as it was also called, as evidence of the fixed place of a species in the natural order of things. This was not an unreasonable assumption for the science of these times, because over the preceding two millennia life had changed little in the eyes of naturalists. Added to this concept was the idea that the Earth was only 6,000 years old (see GEOLOGICAL TIME), which did not allow sufficient time for any meaningful change, other than the minor influences of man in the case of domesticated animals. However, by the 17th century this viewpoint began to change. The most important of these occurred in the study of fossils and the initial thoughts of some scientists that there may be a relationship between the fossil record and the history of life (see PALEONTOLOGY). Coupled to this was the realization that the Earth may be far older than originally thought. In addition, explorations of the globe were revealing that the geographic distribution of life was far more complex than originally thought. In the 18th century some investigators of embryological development began to question the prevailing theory of pre-formation, opting instead for a belief in the gradual development of the embryo over time (see EMBRYOLOGY). As the 19th century began scientists were presented with the possibility of an old Earth in which species may have changed over time. What was lacking was a scientific explanation of the mechanism by which this change occurred.

In the early years of the 19th century several attempts were made to develop theories to explain the process of evolution. One of the first of these was proposed by Jean Baptiste Lamarck, a French naturalist with considerable experience in the study and classification of invertebrate animals. Lamarck's study of the fossil record indicated, as it had to others before him, that some species may have become extinct over time. Lamarck did not favor the idea of extinction, so instead he proposed that organisms had changed over time in response to changes in their environment. His ideas were published between 1809 and 1815 and became known as the theory of acquired characteristics. A prime component of this theory was the belief that individuals could pass acquired characteristics on to the next generation (see THEORY OF ACQUIRED CHARACTERISTICS). Opposing to Lamarck's ideas was the theory of catastrophism, proposed by the French paleontologist Georges Cuvier. Cuvier believed that all organisms had been created at the same time but that a series of catastrophic events had eliminated a number of these species (see PALEONTOLOGY). For several decades the scientific community debated the merits of these two ideas.

In 1831 the British naturalist Charles Darwin embarked on a five-year voyage aboard the *H.M.S. Beagle*, a ship chartered by the English government to chart the coastal areas along the west coast of South America. Darwin was already an accomplished naturalist before his voyage on the *Beagle*, because his grandfather Erasmus Darwin was a naturalist in the early part of the century (see THEORY OF ACQUIRED CHARACTERISTICS). Furthermore he possessed a strong knowledge of geological processes, as well as training in mineralogy and **entomology**. Darwin's knowledge of geology was influenced by Charles Lyell's three-volume series entitled *Principles of Geology*, which he took along with him on the voyage. In this work Lyell made a convincing argument not only that the age of the Earth was far greater than originally thought, but also that the surface of the Earth was dynamic (see GEOLOGY). Darwin observed the dynamic processes of the Earth firsthand while examining the effects of earthquakes in Chile and its ability to shape the height of the shoreline. From his studies of geology he proposed correctly that coral reefs were first formed around islands that had slowly subsided into the sea. Lyell influenced Darwin not only regarding the concept of geology but also the relationship between geological change and the biological sciences. Specifically, he believed that the fossil record served as a record of both the history of life and the geological conditions of the area in which those organisms had lived (see GEOLOGY). Lyell believed that new species were continuously being formed as the Earth changed and that these changes were also responsible for the extinctions apparent from the fossil record.

During the five-year expedition Darwin recorded detailed observations of the diversity of life along the South American coast. It was around this time that he probably began to notice that life was well adapted to its local environment and that although similar organisms could be found around the globe, each possessed minor variations that allowed it to survive in its specific environment. This became especially apparent to Darwin when he visited the Galapagos Islands in 1835. The Galapagos are a chain of volcanic islands located about 600 miles off the coast of Ecuador. There Darwin encountered a much more elaborate display of the adaptation of a species to its environment than he had witnessed along the coast of South America. Of these observations, the most often quoted is the adaptation of the local finch populations to the environmental conditions of the island. On the Galapagos Islands the finches had about the same body size as the mainland finches, but their beaks were specialized for the local food supply. Darwin also encountered an interesting phenomenon with the local populations of giant tortoises. Each island had a unique population of tortoises, which could be distinguished by variations in the shape of their shells. Throughout the journey Darwin witnessed similar geographic variations in nature, and by the end of the journey he was beginning to doubt the idea that all species were created in their current location in favor of the idea of what was called the transmutation of species.

Darwin did not develop his theory of natural selection while on board the *Beagle*, but the scientific data he accumulated on the trip had a major influence on his thinking for some time after his return. Around 1838 Darwin read a book by the English economist Thomas Malthus (1766–1834) that addressed the issue of human resource use. Malthus recognized that although human populations multiply at a geometric rate, the resources to support the populations do not increase at the same pace. At some point, Malthus suggested, natural selection events such as war or famine reduce the population by allowing only the fittest to survive. The concept of the survival of the fittest had been recognized and documented by a large number of naturalists prior to Darwin, but for Darwin the idea of selection provided a focusing point for his developing ideas on how species change over time. Darwin appears to have been unaware of Gregor Mendel's work on genetics, which would have provided useful information about the mechanism of change (see INHERI-TANCE). For the next several years Darwin worked on other projects, but from his correspondence and notes it appears that the idea of how new species form was never far from his thoughts.

At about the same time the English naturalist Alfred Wallace (1823–1913) was conducting a series of research expeditions in Malaysia and South America. Wallace was interested in the geographic distribution of animal species. Just as Darwin was influenced by the geological work of Lyell, Wallace's ideas of evolution were partially shaped by the paleontological studies of Francois Pictet (see PALEONTOLOGY), who believed in a gradual change in species over time. It appears that Wallace was also familiar with Malthus's book and recognized, just as Darwin did, that the concept of selection was important in the descent of animals over time. By the early 1850s both Wallace and Darwin were working independently on the mechanism for evolutionary change. This is not simply a coincidence of history, but rather reflects that a substantial amount of information had been accumulated by the scientific community by the mid-19th century and that the time was ideal for the development of the idea of natural selection. By most accounts Wallace was the first to describe the importance of variation to the change of a species over time and sent copies of his manuscripts to Darwin. The ideas of the two scientists were similar, and they presented simultaneous papers on their ideas at a meeting in 1858. In 1859 Darwin published his landmark book entitled *The Origin of Species by Means of Natural Selection*, which outlined the theory of natural selection (see Figure 38).

This theory combined in one place Darwin's interpretation of the work done by Lamarck, Wallace, Lyell and Malthus. The basic premises of this theory is that the individuals of a population produce more offspring than can survive and thus competition results for available resources. However, the observed variation in the population gives some individuals a selective advantage in this struggle. Based on the basic principles of survival of the fittest, those individuals that were best adapted would survive and thus would pass these traits on to the next generation. Over time the population would change

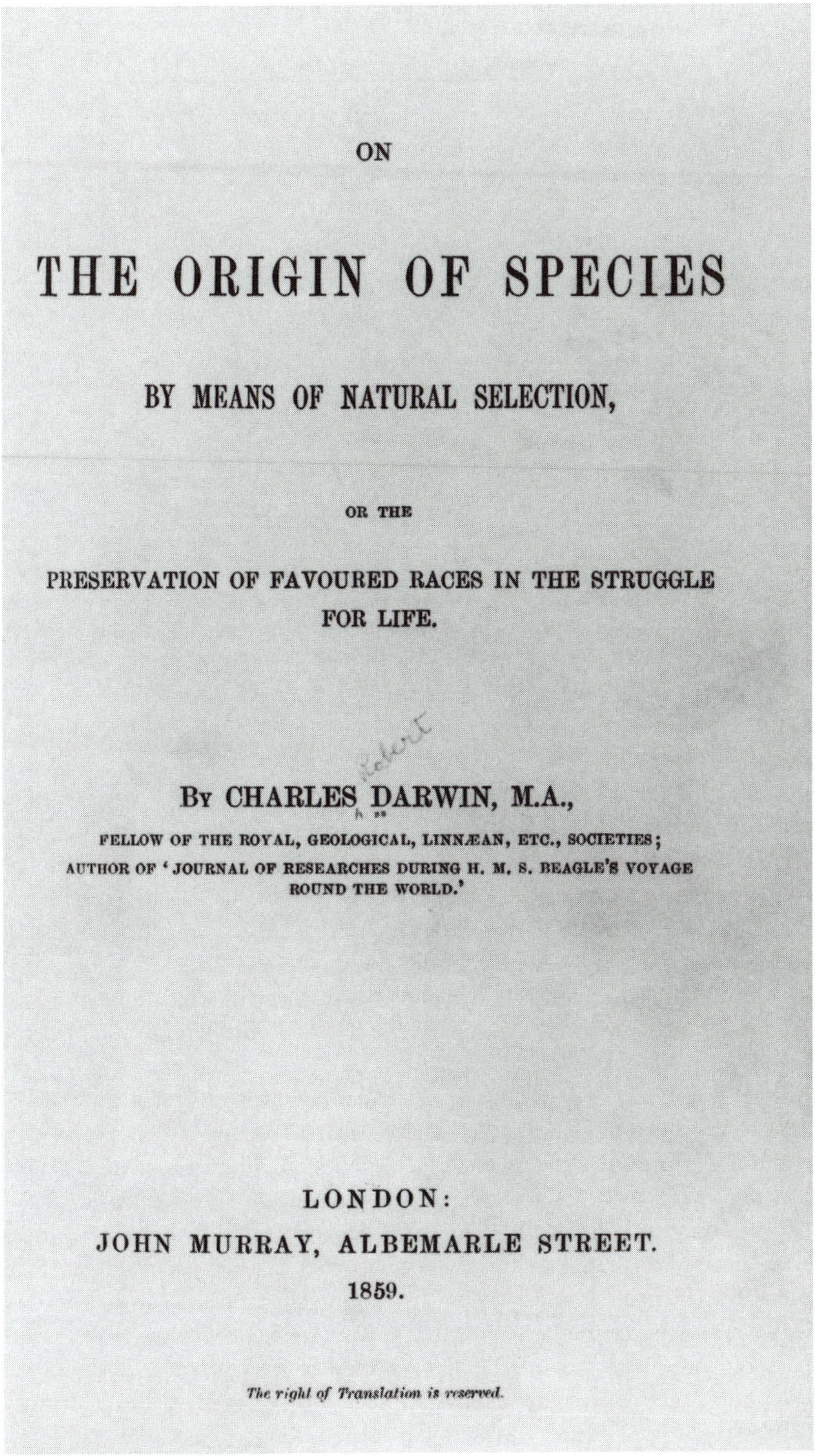

ON

THE ORIGIN OF SPECIES

BY MEANS OF NATURAL SELECTION,

OR THE

PRESERVATION OF FAVOURED RACES IN THE STRUGGLE FOR LIFE.

By CHARLES DARWIN, M.A.,

FELLOW OF THE ROYAL, GEOLOGICAL, LINNÆAN, ETC., SOCIETIES;
AUTHOR OF 'JOURNAL OF RESEARCHES DURING H. M. S. BEAGLE'S VOYAGE
ROUND THE WORLD.'

LONDON:
JOHN MURRAY, ALBEMARLE STREET.
1859.

The right of Translation is reserved.

Figure 38. The cover of Charles Darwin's *On The Origin of Species*, published in 1859. (Library of Congress photo collection)

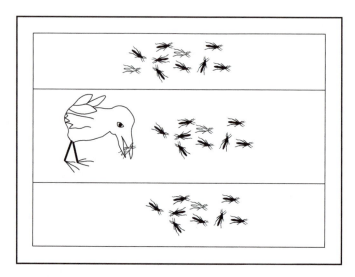

Figure 39. An illustration of the process of natural selection in action. The top diagram shows a natural variation in the color of the insects. If a predator, indicated by the bird, prefers to eat the light-colored insect (middle panel) then the frequency of light-colored insects in the next generation (lower panel) will be lower.

to reflect the advantage of this trait (see Figure 39). Darwin contended that this process of natural selection provided the mechanism for evolutionary change over time. Although in many ways this theory may appear to be similar to Lamarck's early theories (see **THEORY OF ACQUIRED CHARAC-TERISTICS**), Darwin had demonstrated that selection acts on existing variation within a population and did not create the variation as proposed by Lamarck. Furthermore, Darwin correctly deduced that the population was the fundamental unit of evolutionary change, and not the individual as proposed by Lamarck.

The publication of *The Origin of Species* immediately created a controversy among the scientific community. Some, such as Darwin's chief supporter Thomas H. Huxley, viewed natural selection as the most logical and simple mechanism of evolutionary change. Others did not support the idea that natural selection could explain the formation of new species. A major milestone in the acceptance of this theory occurred in 1860 when the German naturalist Hermann von Meyer unearthed a fossilized feather from a quarry in Bavaria. The feather was from a reptile named *Archeopteryx* and the supporters of evolution contested that it represented a transitional species in the evolution of birds from reptiles. Additional studies of horse fossils further validated Darwin's ideas of evolutionary decent.

The controversy over the theory of natural selection was not limited to the scientific community. The Catholic religion, and many scientists, objected to the idea that Darwin's common descent might mean that humans were descended from nonhuman animals, a conclusion that Darwin never directly states in *The Origin of Species*. The supporters of evolution vigorously defended their new theory and firmly established the principle that any legitimate challenge to their theory would have to be based on scientific evidence. The question of the relationship between the theory of natural selection and the evolutionary history of humans continued for the remainder of the century (see HUMAN EVOLUTION) and prompted Darwin to publish *The Descent of Man, and the Selection in Relation to Sex* (1871). The debate between science and religion continues to this day, with new ideas such as scientific creationism and intelligent design continuing to test the foundations of evolutionary science.

Although the publication of *The Origin of Species* may be regarded as the final triumph of evolutionary science, in reality it was simply the start of a debate among the scientific community that continues today. Modern evolutionary biologists now have examples of natural selection in action, as is demonstrated by insecticide resistance in many species of insects, and resistance to antibiotics by many pathogenic microbes. Furthermore, geneticists now recognize that Darwin's theory is based on genetic principles first established by Mendel (see INHERITANCE). The changes that Darwin refers to are actually the result of changes in the DNA of the organism (see BIO-MOLECULES, INHERITANCE) The link between genetics and natural selection was established between 1930 and 1950 in what is now commonly called the modern synthesis. For example, in the case of insecticide resistance, the selective agent (the insecticide) actually selects for variation of genes, called alleles, some of which confer resistance. Evolutionary biologists now recognize two different forms of evolution. First, microevolution represents the change in the allele frequency, or genetics, of an organism over short periods of time. On the other hand, macroevolution is the study of evolutionary principles over geological time spans. Questions remain on both of these ideas, such as the pace of evolution over time. Many paleontologists now discount the idea of gradual evolution and instead recognize that the fossil record actually represents a more sporadic nature to evolution, also called **punctuated equilibrium**. Other biologists now suggest that for some organisms it may be possible that individuals evolve in a manner first suggested by Lamarck (see THEORY OF ACQUIRED CHARACTERISTICS). What this indicates is that the theories presented by Darwin and Wallace in the 19th century have continued to evolve over time and will continue to do so in the future, as science develops a deeper understanding of natural forces.

Selected Bibliography

Bowler, Peter. *Evolution: The History of an Idea*. Berkeley: University of California Press, 1989.
Darwin, Charles. *The Origin of Species*. Baltimore, MD: Penguin Books, 1968.
Mayr, Ernst. *The Growth of Biological Thought: Diversity, Evolution and Inheritance*. Cambridge, MA: Harvard University Press, 1982.
Young, David. *The Discovery of Evolution*. Cambridge, England: Cambridge University Press, 1992.

Typewriter (1868–1895): Along with the printing press, the typewriter represents an important level of automation for the printing industry. While the printing press is designed for the mass production of literature (see PRINTING PRESS), the typewriter is more of a personal instrument. The typewriters of the 19th century would appear very foreign in comparison to modern machines, but they both possess many of the same features.

The invention of the first typewriter is credited to the American inventor Christopher Sholes (1819–1890), but in fact there were attempts earlier in the 19th century to invent an automated system of writing, although none of them met with much success. However, Sholes had some previous experience with the development of a page-numbering machine and was trained as an editor and printer. In 1868 Sholes and two other inventors developed the first version of their typewriter. These machines are commonly called blind typewriters because the user of the device could not view the typed page directly while typing. Instead the typist had to stop and lift a cover to view the typed page. As the operator pressed a key a long type bar swung forward and made contact with a ribbon that contained ink. Each letter had its own bar that was connected to the key by a small wire. The bars themselves were arranged on a disk inside the machine. The first of these typewriters also used a foot pedal, similar to that found in a piano or sewing machine, for the carriage return–the equivalent of the return key on the modern computer.

One of the things that Sholes quickly noted during test runs was that for a rapid typist the circular keys jammed together as they moved forward to make contact with the paper. Sholes and his fellow inventors tried several options, including placing a single row of 44 keys in alphabetical order. However, the solution that Sholes settled on simply separated the type bars of commonly used letter pairs, such as *th* and *st*, so that the keys would not jam as they returned after contacting the paper. When Sholes connected the internal type bars to the keyboard the result was what appeared to be a jumbled mass of letters. Sholes's configuration of letters is commonly called the QWERTY keyboard, after the first six letters from left on the top row of the keys. These first typewriters did not get much attention from the business community. Even authors and journalists did not see merit in the first versions of the typewriter. The first book that was believed to have been authored using a typewriter was prepared by Mark Twain in 1883. After encountering a less than responsive market Sholes sold the patent to the Remington Arms Company in 1874, which immediately began a more intensive marketing campaign. By the start of the 20th century the use of the typewriter was becoming more widespread.

There have been a number of improvements to the Sholes machine over the past century. At the very end of the century (1895) the first typewriter was developed that allowed the operator to view the printed page as it was being typed. Figure 40 shows an example of a late-19th century typewriter. In the 20th century electric typewriters were introduced as were word processing machines, which were early versions of the modern computer. Some attempts

The First Machine to get into General Use.

Figure 40. An example of a late-19th century typewriter. (Hulton Archive/Getty Images)

have been made to develop a more logical keyboard that makes better arrangement of the letters, such as the 1930-era Dvorak keyboard, but none of these have gained widespread popularity.

The invention of the typewriter is significant for its impact on society as a whole. While the earliest versions of the machine were purchased more out of curiosity than for use as a timesaving device, the typewriter became a standard instrument in the 20th century workplace. Despite having survived for more than a century, the standard typewriter is rapidly being replaced by computers and inexpensive printing methods that allow greater versatility for the user. Typing on a computer keyboard allows for corrections and revisions that are difficult to make using traditional typewriters, and the modern printer allows multiple copies to be made almost instantly. Although the typewriter itself may someday become a casualty of technological advance, a reminder of its past dominance will remain in the QWERTY keyboard, first developed during the 19th century.

Selected Bibliography

Karwatka, Dennis. *Technology's Past*. Ann Arbor, MI: Prakken Publications, 1996.

Rehrer, Darryl. "The Typewriter." *Popular Mechanics* 173, no. 8 (1996): 56–59.

Van Dulken, Stephen. *Inventing the 19th Century: 100 Inventions That Shaped the Victorian Age from Aspirin to the Zeppelin*. New York: New York University Press, 2001.

V

Vaccinations (1796–ca. 1890): A vaccination is a medical procedure in which a small amount of a disease-causing agent, such as a **virus**, is
delivered to an individual in attempt to enhance the immune response against
the disease. Attempts to control the spread of a specific disease by exposing
individuals to a related type of disease had been recorded by the ancient
Chinese, Indian and Persian civilizations. In the case of smallpox the Chinese
frequently used the dried scabs of the disease to protect citizens from the
active form of the disease. For some time there had been an understanding
in Western civilization that some smallpox outbreaks were less virulent than
others, and it was recognized that once someone contracted one of these less
destructive strains they were immune to further infection. However, it was not
until the late 18th century that the first serious scientific attempt was made to
use a vaccination to control a disease in Western society. In 1796 the English
physician Edward Jenner (1749–1823) was investigating a prevention for
smallpox used by farmers in rural England in which people were first exposed
to a similar disease called cowpox (see Figure 41). It was believed by the local
farmers that once a person had been exposed to cowpox they were immune
to smallpox, a far more serious disease with a mortality rate in excess of 30%.
Jenner tested the validity of this idea by intentionally injecting an 8-year-old
boy with cowpox. Approximately 2 months later he injected the same boy
with smallpox. However the boy did not develop any smallpox symptoms.
Similar experiments with additional patients yielded the same results (see
Figure 42). Over the next several decades this type of smallpox vaccination
was responsible for limiting the spread of the disease in many areas of
England. Jenner is credited with developing the word *vaccination*, from the
Latin word for "cowpox," to describe this type of procedure. In the 19th
century research into this area of medicine would be greatly expanded, mostly

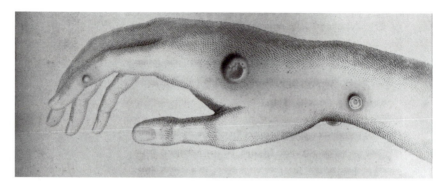

Figure 41. An example of cowpox, which was initially used to vaccinate patients against smallpox. (Library of Congress photo collection)

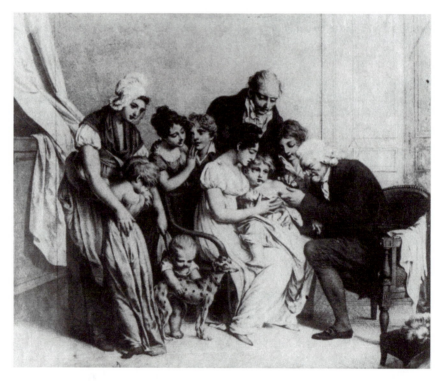

Figure 42. An illustration of Jenner vaccinating his son in the 19th century. Within a few years this procedure was being used across England. (Hulton Archive/Getty Images)

because of the widespread recognition that microscopic organisms were responsible for many forms of disease (see GERM THEORY).

One of the leaders in the development of vaccinations in Europe during the 19th century was the French chemist Louis Pasteur. Pasteur had gained a reputation as a skilled scientist after solving problems of both the wine and silk industries in France. In his studies he established many of the fundamental principles about how microbial organisms interact with the process of fermentation (see FERMENTATION) and act as disease-causing agents (see GERM THEORY). When Pasteur's ideas were combined with the innovative scientific procedures of Robert Koch, whose work defined how scientists establish the relationship between suspect microorganisms and disease (see GERM THEORY), the stage was set for one of the most important advances in the study of medicine. As is frequently the case in science, Pasteur's first scientific breakthrough in the study of vaccinations occurred by accident. Following his earlier successes with the germ theory of disease, Pasteur was intensely interested in isolating the agents responsible for other common diseases. One of these diseases was cholera, which is responsible for many forms of **dysentery**. Pasteur was working with a highly lethal form of cholera in chickens, and on one occasion he mistakenly injected a group of chickens with a strain of cholera that had been prepared some time in the past. Whereas the cholera strain he was working with was normally lethal, this time the chickens displayed only minor symptoms of the disease and did not die. Furthermore, when he subsequently injected the same chickens with the potent strain of the disease, they showed no signs of contracting cholera. What Pasteur had discovered was the concept of *attenuation*, in which the potency of a disease-causing agent is reduced so that it will produce an immune response but not sickness. Pasteur used the same approach in his studies of the disease anthrax. Anthrax was a major problem of the farm industry, where it infected both farm animals and their human caretakers. In 1876 Robert Koch had isolated the bacterium responsible for anthrax (see GERM THEORY) and by 1881 Pasteur had developed an attenuated strain of the bacteria that could be used to vaccinate both animals and humans against the disease.

To this point Pasteur's work had involved developing vaccinations for diseases that had microbial causes. However, around 1885 Pasteur began work on a disease called rabies. Rabies was a challenge for Pasteur because he was unable to isolate the bacterium responsible for causing the disease. It is now recognized that rabies is the result of a virus, but it was not until 1892 that Dmitri Ivanovski (1864–1920) and later Martinus Beijerinck (1851–1931) started accumulating evidence that these infectious agents existed. Since Pasteur was unable to isolate the bacterium for cultivation, he used rabbit hosts to generate material to produce his vaccinations. Once again he attenuated the potency of the vaccination by aging the infected material before injecting it into the patient. In 1885 Pasteur was successful in using the vaccine to prevent a case of rabies in a small boy suspected of being bit by a rabid dog.

As the 19th century drew to a close there was one additional important advance in the development of vaccines. One of the prime problems with Pasteur's method was that assessing the potency of the attenuated vaccine was exceedingly difficult without injecting it into a test subject. Because individual immune responses vary, there was always the chance that the vaccination may itself cause the disease. In 1890 a partial solution to this problem was provided by Emil von Behring (1854–1917). Instead of injecting attenuated material directly into the patient to provide immunity, von Behring first injected the material into an animal and then isolated the animal's defensive compounds. This material, called an antitoxin, could then be injected safely into a patient without risking infecting the patient with the disease.

The fact that a vaccination allows the immune system of the patient to produce defensive proteins called antibodies was not known until the 1920s. Since that time significant advances have been made in the development of vaccines, but the fundamental principles of isolating the disease-causing agent and preparing a vaccine have changed little since the work of Koch and Pasteur. The development of vaccines now involves the work of chemists, molecular biologists and virologists, scientists who specialize in the study of viruses. Regular vaccinations, especially in the early years of life, have been credited with greatly expanding our life span and the health of our civilization. Many diseases, such as smallpox, have effectively been eradicated by the use of vaccines, and scientists now think that it may be possible in the near future to develop vaccines against cancer cells, thus allowing our bodies to fight both internal and external enemies. However, widespread vaccination programs are not sufficient to eliminate all threats of a specific disease. Although smallpox was effectively eliminated in the 1970s, a number of countries have retained stockpiles of the disease for use as a biological weapon in a time of war. For this reason there has been renewed interest in assessing vaccination protocols in the case of a biological attack and the development of new vaccines against possible mutated stocks. For example, the United States government has recently requested more than 300 million smallpox vaccines and is currently developing anthrax vaccines as a protection against terrorist attacks. The success of vaccines and vaccination programs may be attributed to the pioneering efforts of a few scientists in the 19th century.

Selected Bibliography

Asimov, Isaac. *Asimov's New Guide to Science*. New York: Basic Books, 1984.

Duffin, Jacalyn. *History of Medicine: A Scandalously Short Introduction*. Toronto: University of Toronto Press, 1999.

Pfeiffer, Carl J. *The Art and Practice of Western Medicine in the Early Nineteenth Century*. Jefferson, NC: McFarland, 1985.

APPENDIX: ENTRIES BY SCIENTIFIC FIELD

Astronomy

Asteroids

Astronomy

Earth

Photography

Planetary Astronomy

Spectroscope

Sun

Biology

Biomolecules

Cell Biology

Cell Division

Cell Theory

Embryology

Enzymes

Fermentation

Germ Theory

Human Evolution

Inheritance

Organic Chemistry

Paleontology

Synthetic Dyes

Theory of Acquired Characteristics

Theory of Natural Selection

Vaccinations

Chemistry

Atomic Theory

Atomic Weight

Biomolecules

Cathode-Ray Tube

Chemistry

Elements

Fermentation

Gas Laws

Kinetic Theory of Gases

Nitroglycerine

Organic Chemistry

Spectroscope

Geology

Earth

Fossil Fuels

Geological Time

GLOSSARY OF
TECHNICAL TERMS

Absolute Temperature (Chemistry, Physics). A temperature measured from absolute zero, or zero degrees on the Kelvin scale.

Acid Rain (Chemistry). Any form of precipitation that has a pH below 5.6. It is usually caused by atmospheric pollution by sulfuric or nitric acids.

Alchemy (Chemistry). This precursor to modern chemistry focused on the transmutation of matter, specifically the procedures to convert certain metals into gold.

Alleles (Biology). In genetic studies these are the variations of a specific gene. Alleles may cause variations in the phenotype, or characteristics, of the organism and are the result of variations at the DNA level.

Alternating Current (Physics). An electrical current in which the flow of electricity alternates direction from positive to negative over time, as is indicated by a wave. It is frequently abbreviated as AC.

Anaerobic (Biology, Chemistry). Describes a biochemical reaction that occurs in the absence of oxygen. When used to describe an organism it indicates the organism's ability to live without oxygen.

Analytical Geometry (Mathematics). The definition of geometric shapes and curves using algebraic functions. It involves the plotting of the shape on a coordinate system to define its physical properties.

Anode (Physics). The positive terminal of a battery.

Arthropods (Biology). The phylum of the animal kingdom containing the largest number of species, with more than a million currently identified. They are characterized as having jointed appendages, a protective exoskeleton and a developed body cavity. Insects are the largest group of arthropods.

Atomic Weight (Chemistry). A unit used in chemistry to indicate the weight of a single atom of an element. Because there may be some variation in the weight, periodic tables often indicate the atom's average weight in nature.

Axioms (Mathematics). The assumptions or truths on which a mathematical theory is based.

Barometric Pressure (Meteorology). The pressure of the atmosphere on the surface of the earth, measured as the height that a volume of mercury is elevated within a tube. Units are typically in millibars, although inches may still be used at times.

Binary Fission (Biology). A type of asexual division in bacteria that involves the division of the genetic material. The cells produced by this procedure are usually genetically the same.

Calcite (Geology). A common mineral frequently found in limestone. It has the chemical formula $CaCO_3$ and is sometimes called Iceland spar after a region in Iceland that contains large crystals of this mineral.

Calculus (Mathematics). A branch of mathematics that studies the rate of change.

Calorie (Physics, Biology). The amount of energy required to raise 1 gram of pure water from 15°C to 16°C at sea level. The joule is now frequently used by the scientific community.

Cancer (Biology). The unregulated growth of cells within an organism, frequently caused by genetic or metabolic changes at the cellular level.

Capacitor (Physics). An electrical device that holds a charge by separating two conductors with some form of insulating material (air, plastic, glass). Capacitors are usually designed for a rapid discharge of current when triggered.

Carbon Dating (Biology). A method of dating organic material using radioactive carbon molecules (carbon-14 [^{14}C]). Living organisms acquire ^{14}C at a specific rate from the atmosphere. A ratio of ^{14}C to nonradioactive ^{12}C may be used to indicate when the tissue or organism stopped metabolic functions or died.

Cataracts (Medicine). A medical condition associated with the cornea of the eye and caused by a clouding of the normally translucent cells. This is frequently an indication of additional medical problems such as diabetes.

Cathode (Physics). The negative end or terminal of a battery.

Cellular Respiration (Biology). The biological process by which a cell releases the energy of organic molecules, such as sugar, to provided for the metabolic needs of the cell. If this process requires oxygen, it is also called aerobic respiration.

Centrifugal Force (Physics). For an object on a curved path, this is the outward force directed away from the center of motion. This force would attempt to return the object to motion in a straight line.

Chordate (Biology). A phylum of the animal kingdom that is characterized as having a dorsal notochord, a postanal tail, pharyngeal slits and a dorsal nerve cord. This phylum consists of both vertebrate and invertebrate species and is the phylum of *Homo sapiens*.

Combustion (Chemistry). The release of heat from a substance as a result of burning.

Commutative Law of Multiplication (Mathematics). A mathematical law that states that in addition and multiplication the order of the quantities does not affect the sum or product of the equation.

Condensation Reaction (Biology, Chemistry). A chemical reaction in which water is removed to form a chemical bond.

Cytoplasm (Biology). The aqueous region contained by the cell membrane. It is also sometimes called the cytosol.

Cytoskeleton (Biology). The dynamic internal framework of the cell. It is composed of protein structures called microtubules, microfilaments and intermediate filaments.

Density (Chemistry). The ratio of the mass of a substance to its volume. It is usually expressed as grams per liter or an equivalent measurement.

Diatoms (Biology). Single-celled, photosynthetic, aquatic organisms that possess a silica covering. They belong to the kingdom Protista. More than 5,500 living species are known.

Diffraction (Physics). The bending of a light wave from its original straight-line path.

Distillation (Chemistry). A chemical process that is used to separate a solution into its components. It involves heating the mixture until one of the components evaporates and then condensing the vapor on a cooler surface. Most commonly used in the beverage industry to remove alcohol from fermenting organic material.

Double Refraction (Physics). The splitting of a beam of light into two components, each of which travels in a different direction. Not all materials are capable of double refraction, but some minerals are. The process is sometimes called birefringence.

Dysentery (Biology). A medical condition involving inflammation of the large intestine that is characterized by bleeding and diarrhea. It is usually caused by an infection of an amoeba or bacteria.

Echinoderms (Biology). A phylum of living organisms that is characterized as marine invertebrates. An example is the sea star.

Electrochemistry (Chemistry). A branch of chemistry that focuses on the interaction of electricity and chemistry and how different materials react to electrical charges.

Electrolyte (Physics). In the study of electricity, any material that moves an electrical current using ions rather than electrons. Electrolytes are usually a liquid or gas, as opposed to solids that move current via electrons.

Empirical (Chemistry). A method of indicating the proportion of elements in a given compound. It differs from the molecular formula, which lists the actual number of atoms for each element in the compound.

Entomology (Biology). The branch of the biological sciences that specializes in the study of insects.

Entropy (Physics). The measure of disorder in a system, usually used in the study of thermodynamics.

Equatorial Mount (Astronomy). A type of mounting for a telescope that uses two perpendicular axes, a polar axis and declination axis, to orientate the instrument.

Erg (Physics). A unit of energy associated with the centimeter-gram-second system of units. It is equal to approximately 10^{-7} joules

Ether (Chemistry). Before the 20th century, ether was considered to be the element or medium that filled the space between atoms. The modern term defines a specific chemical compound containing oxygen. It belongs to a class of compounds called the alkoxyalkanes.

Eukaryotic (Biology). A type of cell that is characterized as having a nucleus and internal membrane-bound organelles.

Forensic Medicine (Biology, Medicine). The application of medical evidence to criminal court cases. Forensics indicates a process associated with law.

Frequency (Physics). The rate of repetition of a wave. If it is an electromagnetic wave, this rate is typically measured in wavelengths per second or hertz.

Functional Groups (Chemistry). The portion of a chemical compound that is defined as having specific chemical characteristics. An example would be hydroxyl (OH-) or carboxyl (COOH) groups.

Genes (Biology). The fundamental unit of inheritance. It is also regarded as the region of the genetic material that provides the instructions for the construction of a specific protein in living cells.

Genome (Biology). The total genetic content of an organism or virus.

Glaucoma (Medicine). A group of eye diseases that cause damage to the optic nerve that delivers vision messages to the brain.

Glycerine (Chemistry). A chemical compound, also called glycerol, that is derived from the production of soap. It has the chemical formula $C_3H_5(OH)_3$.

Histology (Biology). The study of plant and animal tissues, usually focusing on their structure and chemical composition.

Horsepower (Physics). A unit of power in the English system of measurement. It is frequently used to indicate the output of motorized equipment and is equal to about 747.5 watts.

Hydrocarbons (Chemistry). A class of chemical compounds that consist of only carbon and hydrogen.

Hydrologic Cycle (Geology). Also called the water cycle, this is the full range of water movement on the planet or in an ecosystem. It involves not only bodies of water, but also the atmosphere and subterranean sources of water.

Hydrophobic (Chemistry). A class of molecules that do not interact with water because of their chemical composition. An example is the lipids.

Hyperbolic (Mathematics). Associated with the study of the hyperbola, which is a geometric shape that is formed by intersecting a circular cone with a plane parallel to the axis of the cone.

Inclination (Astronomy). The angle between the ecliptic and the orbital planes of a planet or satellite. The deviation from the normal orbital plane of the solar system.

Intelligent Design (Biology). A form of theological thinking that acknowledges evolution but insists that evolutionary processes have been guided by a divine being.

Inverse-Square Law (Physics). Any law in which the physical quantity being measured varies with the distance from the source inversely as the square of that distance. Gravity is frequently defined by the inverse-square law.

Invertebrate (Biology). A term used to describe organisms that lack a backbone or other internal skeleton. Of all living creatures, 99% are invertebrates.

In Vitro (Biology). A biological reaction that occurs outside the cell in an artificial environment such as a test tube.

Ionic (Chemistry). A type of chemical reaction that involves ions, which are electrically charged atoms or molecules.

Irrational (Mathematics). A type of mathematics in which a number is not able to be expressed as the ratio of two integers.

Isotopes (Chemistry). Two or more atoms that have the same number of protons (atomic number) but different numbers of neutrons.

Joules (Physics). A unit of energy that is used in the meter-kilogram-second system of measuring. It is equal to one watt-second of energy.

Laser-Argon Dating (Geology). Also known as potassium-argon dating. A method of dating geological material using the decay rate of radioactive potassium. Radioactive potassium decays to form an isotope of argon that is released with the use of lasers. The ratio of the potassium to argon gives an approximation of the geological age of the material being tested.

Light-year (Astronomy). The distance that light travels in one year. It is approximated as 9.461×10^{12} kilometers.

Logic (Mathematics). The study of the principles of reasoning. Also used in the design of electronic and computer circuits to indicate yes-or-no type decisions.

Matter (Chemistry). Anything that has mass and occupies space.

Micrometer (Astronomy). A mechanical device used in the 17th century to measure the distances between stars against the backdrop of the night sky.

Microwave (Physics). An electromagnetic wave that has a wavelength between 0.3 and 30 centimeters. When exposed to microwave radiation, water oscillates, causing an increase in temperature.

Molecular Biology (Biology). The area of the biological sciences that addresses life at the molecular level. It frequently includes aspects of cell biology and molecular genetics.

Mollusk (Biology). A diverse phylum of the animal kingdom whose organisms are considered to be the first to have evolved a true body cavity. It is the second largest phylum (after Arthropods) and contains an estimated 85,000 living and extinct species.

Mutation (Biology). An unintentional change in the genetic material. Biologists consider mutations to be the source of all genetic variation.

Natural Logarithms (Mathematics). The study of logarithms that use as the base the number *e*. They are also called Napierian logarithms.

Notochord (Biology). A tough dorsal support rod that is present at some point in the lifecycle of all chordate animals. It is the precursor of the vertebral column.

Objective Lens (Biology). In a compound microscope this lens performs the majority of the magnification. Objective lenses are typically closest to the object being viewed and may be supplemented by additional magnification in the eyepiece (ocular lens) of the instrument.

Oscilloscope (Physics). An instrument used to analyze electrical signals. It uses a cathode-ray tube that projects onto a fluorescent screen.

Pasteurization (Biology). A process by which microbes are removed from a solution by raising the temperature of the liquid for an extended time.

Pathology (Biology, Medicine). The study of the causes of diseases and their effects on the host organism.

Pedigree (Biology). In genetic analysis this is the genetic history of an individual traced back through its parents for several generations.

Phenotype (Biology). In the study of genetics this is the observable characteristics of an organism. The genetic makeup of an organism, or its combination of alleles, is called its genotype.

Phlogiston (Chemistry). A hypothetical substance that was used to explain the process of combustion. Phlogiston was believed to be colorless, odorless and weightless. For a brief time hydrogen was believed to be the equivalent of phlogiston.

Photometer (Astronomy). An instrument that measures the portion of the electro-magnetic spectrum that is associated with the visible wavelengths of light.

Pi (Mathematics). The ratio of the circumference of any circle to its radius. It is an irra-tional number that is estimated at 3.14159.

Piezoelectricity (Physics). The conversion of mechanical force to electrical force, or vice versa. Some crystals can convert electrical input into sound waves.

Planck's Constant (Physics). A physical constant that can be defined as the ratio of the energy of a photon (light particle) to its frequency. It is sometimes called the quantum of action in the study of quantum physics.

Plate Tectonics (Geology). The study of the movement of the plates of the earth's crust over time. It also involves the study of volcanoes, earthquakes and faults associated with these plates.

Platelets (Biology). Also called thrombocytes, these small structures of the blood assist in the formation of a blood clot.

Population (Biology). A group of interbreeding organisms that occupy a given geo-graphical region.

Population Genetics (Biology). The study of patterns of inheritance at the popula-tion level, including the frequency of alleles in a population and its variance over time.

Porifera (Biology). A phylum of the animal kingdom best represented by the sponges. These organisms lack tissues and organs and are considered to be the sim-plest of the animals. They are sometimes placed in their own subkingdom called the Parazoa.

Potential Energy (Chemistry, Physics). The energy stored in the chemical bonds of a compound. It is also defined as the capacity to do work. Once the energy is used to conduct work it is called kinetic energy.

Primates (Biology). An order of mammals in the animal kingdom to which humans, monkeys, apes and lemurs belong.

Prokaryotic (Biology). A group of single-cell organisms that lack membrane-bound organelles or a nucleus. They are considered to be some of the oldest organisms on the planet. Bacteria are prokaryotic.

Protistans (Biology). A kingdom of life that consists of single-cell eukaryotic organisms. The members of the kingdom are typically not related to one another.

Punctuated Equilibrium (Biology). A concept used to explain the pattern of evolution over time. As opposed to gradualist ideas, punctuated equilibrium states that a species may undergo a brief period of rapid evolution or change, followed by an extended period of little or no change.

Quinine (Medicine). A chemical derived naturally from the bark of the cinchona tree that is used to treat malaria.

Radiant Heat (Physics). Heat that is emitted from an object due to the laws of thermodynamics.

Recombinant DNA (Biology). The manipulation of genetic material (DNA) in the laboratory to produce new combinations of genetic traits.

Refraction (Physics). In the study of electromagnetic radiation, the change in direction of any electromagnetic wave, including light. It is usually caused as the wave changes from one medium to another.

Reye Syndrome (Medicine). A childhood disease characterized by a breakdown of the liver. The exact cause is unknown, but it is believed to be associated with the combination of a viral factor and the use of aspirin.

Rickets (Medicine). A disease that is characterized by deformities of the bone caused by insufficient vitamin D in the diet.

Scientific Creationism (Biology). Also called creation science, it is a belief in the biblical chronology of the creation of the world. However, because this theory is not testable by the scientific method, the scientific community does not recognize it as a science.

Sex-linked Traits (Biology). Traits that are passed on to the next generation on the sex chromosomes of the species. For this reason, the expression of these traits does not adhere to Mendelian laws of inheritance.

Species (Biology). There are a number of definitions of a species, but the most widely accepted is: a group of naturally occurring organisms that have the capability to interbreed. This is the lowest level of taxonomic classification.

Specific Heat (Physics). Usually expressed as a ratio, this is the amount of heat required to raise 1 gram of a given substance 1 degree Celsius, compared to the amount of heat required to raise 1 gram of water the same amount under constant pressure.

Spectral Analysis (Physics). The study of the electromagnetic spectrum of a atom, compound or object. It may examine the wavelengths either emitted or absorbed by the object and usually is performed with the use of an instrument called a spectroscope.

Stoichiometry (Chemistry). The study of the numerical relationships of the products and reactants in a chemical reaction.

Tetrahedral (Chemistry). A three-dimensional geometric object that has four faces, best represented by a triangular pyramid.

Theorem (Mathematics). A statement of mathematical truth that is presented along with its qualifying conditions.

Thermal Radiation (Physics). Any energy that is radiated from liquids, solids or gases as a result of their temperature.

Thermochemistry (Chemistry, Physics). The study of the movement of heat during chemical reactions and changes in the state, or phase, of the matter.

Trade Winds (Meteorology). Patterns of air movement that occur between 25 and 30 degrees both north and south of the equator. They are associated with the movement of the earth on its axis.

Transgenic (Biology). Any organism that contains the genetic material (DNA) of an unrelated organism in its cells. Transgenic organisms are commonly used in agricultural practices to produce disease-resistant crops, although the procedure can be performed in almost any species.

Transmutation (Chemistry). The changing of one element into another. In modern chemistry this is known to involve changes in the atomic structure, specifically the number of protons in the nucleus.

Vacuoles (Biology). Membrane-bound structures within a cell that act as either storage or the site of chemical reactions.

Valence (Chemistry). A number that represents the number of chemical bonds an element will form with other elements. The number is limited by the configuration of the valence electrons in the outermost energy shell of the atom.

Vector (Mathematics, Physics). Any quantity that has both magnitude and direction.

Vertebrate (Biology). A group of the phylum of chordates in the animal kingdom that is characterized as having a backbone that surrounds the nerve chord. Fish, amphibians, reptiles, birds and mammals are all vertebrates.

Virus (Biology). A nonliving infectious agent that consists of a protein covering containing a small piece of genetic material. The genetic material may be either DNA or RNA. Viruses lack metabolic processes and must use the host's cellular machinery to reproduce.

Watts (Physics). A unit of power equal to one joule per second and commonly used as a measurement for electrical devices.

BIBLIOGRAPHY

Albritton, Claude C. Jr. *The Abyss of Time: Changing Conceptions of the Earth's Antiquity after the Sixteenth Century*. San Francisco, CA: Freeman, Cooper, 1980.

Allen, Garland, and Jeffrey Baker. *Biology: Scientific Process and Social Issues*. Bethesda, MD: Fitzgerald Science Press, 2001.

Archer, Gleason, L. *History of Radio to 1926*. New York: Arno Press, 1971.

Asimov, Isaac. *Asimov's Biographical Encyclopedia of Science and Technology*. Garden City, NY: Doubleday, 1972.

———. *Asimov's New Guide to Science*. New York: Basic Books, 1984.

———. *Asimov's Chronology of Science and Discovery*. New York: HarperCollins, 1994.

Baker, V. R. Catastrophism and Uniformitarianism: Logical Roots and Current Relevance in Geology. In Baum, Richard, and William Sheehan (eds.). *Lyell: the Past is the Key to the Present*,

———. *In Search of Planet Vulcan: The Ghost in Newton's Clockwork*. New York: Plenum Trade, 1997.

Beckmann, Petr. *A History of Pi*. Boulder, CO: Golem Press, 1971.

Bennett, Charles, H. Demons, Engines and the Second Law. *Scientific American*. (Nov. 1987): 108–116.

Bennion, Elisabeth. *Antique Medical Instruments*. Berkeley: University of California Press, 1979.

Blundell, D. J., and A. C. Scott. London, England: Geological Society Special Publications, 143 (1998): 171–182.

Bordeau, Sanford P. *Volts to Hertz. . . . The Rise of Electricity*. Minneapolis, MN: Burgess Publishing, 1982.

Bowers, Brian. *Lengthening the Day: A History of Lighting Technology*. New York: Oxford University Press, 1998.

Bowler, Peter. *Fossils and Progress*. New York: Science History Publications, 1976.

———. *The Mendelian Revolution: the Emergence of Hereditarian Concepts in Modern Science and Society*. Baltimore: Johns Hopkins University Press, 1989.

———. *Evolution: The History of an Idea*. Berkeley: University of California Press, 1989.

Bromley, Alan G. Difference and Analytical Engines. In Aspray, William (ed.). *Computing Before Computers*. Ames: Iowa State University Press, 1990.

Brooker, Robert J. *Genetics: Analysis and Principles.* Menlo Park, CA: Addison Wesley Longman, 1999.

Brown, G. I. *The Big Bang: A History of Explosives.* Phoenix, AZ: Sutton Publishing, 1998.

Brush, Stephen G. *The Kind of Motion We Call Heat.* Amsterdam, Netherlands: North-Holland Publishing, 1976.

Bud, Robert, and Deborah Jean Warner (eds.). *Instruments of Science.* New York: Garland Publishing, 1998.

Buffetaut, Eric. *A Short History of Vertebrate Paleontology.* London, England: Croom Helm, 1987.

Burkhardt, Richard W. Jr. *The Spirit of System: Lamarck and Evolutionary Biology.* Cambridge, MA: Harvard University Press, 1995.

Bynum, W. F., E. J. Browne and Roy Porter (eds.). *Dictionary of the History of Science.* Princeton: Princeton University Press, 1981.

Cardwell, D. S. L., *From Watt to Clausius: The Rise of Thermodynamics in the Early Industrial Age.* Ithaca: Cornell University Press, 1971.

A Chronological History of Electrical Development from 600 B.C. New York: National Electrical Manufacturers Association, 1946.

Cobb, Cathy, and Harold Goldwhite. *Creations of Fire: Chemistry's Lively History from Alchemy to the Atomic Age.* New York: Plenum Press, 1995.

Coleman, William. *Biology in the Nineteenth Century: Problems of Form, Function and Transformation.* London, England: Cambridge University Press, 1977.

Concise Dictionary of Scientific Biography, 2nd ed. New York: Charles Scribner's Sons, 2000.

Corsi, Pietro. *The Age of Lamarck: Evolutionary Theories in France 1790–1830.* Berkeley: University of California Press, 1988.

Darwin, Charles. *The Descent of Man and Selection in Relation to Sex.* New York: Hurst , 1874.

———. *The Origin of Species.* Baltimore: Penguin Books, 1968.

Davis, Audrey B. *Medicine and Its Technology: An Introduction to the History of Medical Instrumentation.* Westport, CT: Greenwood Press, 1981.

Dressler, David. *Discovering Enzymes.* New York: Scientific American Library, 1991.

Duffin, Jacalyn. *History of Medicine: A Scandalously Short Introduction.* Toronto, ONT: University of Toronto Press, 1999.

Duke, Martin. *The Development of Medical Techniques and Treatments: from Leeches to Heart Surgery.* Madison, WI: International Universities Press, 1991.

Eicher, Don L. *Geologic Time.* Englewood Cliffs, NJ: Prentice-Hall, 1968.

Fisher, Douglas Alan. *The Epic of Steel.* New York: Harper & Row Publishers, 1963.

Flatow, Ira. *They All Laughed: From Lightbulbs to Lasers, the Fascinating Stories behind the Great Inventions That Have Changed Our Lives.* New York: HarperPerennial, 1993.

Forbes, R. J. *Studies in Early Petroleum History.* Leiden, Netherlands: E. J. Brill, 1958.

Fox, Robert. *The Caloric Theory of Gases from Lavoisier to Regnault.* London, England: Oxford University Press, 1971.

Fradin, Dennis B. *"We have conquered pain": The Discovery of Anesthesia.* New York: Simon and Schuster, 1996.

Francis, Richard L. Did Gauss Discover That, Too? *Mathematics Teacher* 59 (1986): 288–293.

Friedel, Robert, and Paul Israel. *Edison's Electric Light: Biography of an Invention.* New Brunswick, NJ: Rutgers University Press, 1986.

Fruton, Joseph H. *Molecules and Life: Historical Essays on the Interplay of Chemistry and Biology.* New York: John Wiley & Sons, 1972.

Geison, Gerald L. *The Private Science of Louis Pasteur*. Princeton: Princeton University Press, 1995.

Gernsheim, Helmut. *A Concise History of Photography*. London, England: Thames and Hudson Publishers, 1965.

Gernsheim, Helmut. *The Origins of Photography*. London, England: Thames and Hudson Publishers, 1982.

Gillispie, Charles (ed.). *Dictionary of Scientific Biography*. New York: Charles Scribner's Sons, 1970.

Gohau, Gabriel. *A History of Geology*. London, England: Rutgers University Press, 1990.

Gould, Stephen Jay. *Time's Arrow, Time's Cycle: Myth and Metaphor in the Discovery of Geological Time*. Cambridge, MA: Harvard University Press, 1987.

Grattan-Guiness, Ivor. *The Norton History of Mathematical Sciences: The Rainbow of Mathematics*. New York: W. W. Norton, 1997.

Greene, Mott, T. *Geology in the Nineteenth Century*. Ithaca, NY: Cornell University Press, 1982.

Grosvenor, Edwin S., and Morgan Wesson. *Alexander Graham Bell: The Life and Times of the Man Who Invented the Telephone*. New York: Harry N. Abrams Publishers, 1997.

Hardenberg, Horst O. *The Middle Ages of the Internal Combustion Engine: 1994–1886*. Warrendale, PA: Society of Automotive Engineers, 1999.

Haufbauer, Karl. *Exploring the Sun: Solar Science since Galileo*. Baltimore, MD: Johns Hopkins University Press, 1991.

Hessenbruch, Arne (ed.). *Reader's Guide to the History of Science*. London, England: Fitzroy Dearborn Publishers, 2000.

Hills, Richard L. *Power from Steam: A History of the Stationary Steam Engine*. Cambridge, England: Cambridge University Press, 1989.

Hornsby, Jeremy. *The Story of Inventions*. New York: Crescent Books, 1977.

Hoskin, Michael. *The Cambridge Illustrated History of Astronomy*. Cambridge, England: Cambridge University Press, 1997

———. *The Concise History of Astronomy*. New York: Cambridge University Press, 1999.

Humphreys, John. Lamarck and the General Theory of Evolution. *Journal of Biological Education*, 30, no. 4 (1996): 295–304.

Israel, Paul. *Edison: A Life of Invention*. New York: John Wiley & Sons, 1998.

Jack, David B. One Hundred Years of Aspirin. *Lancet* 350, no. 9075 : 437–439.

Karwatka, Dennis. *Technology's Past: America's Industrial Revolution and the People Who Delivered the Goods*. Ann Arbor, MI: Prakken Publications, 1996.

———. *Technology's Past, Vol. 2: More Heroes of Invention and Innovation*. Ann Arbor, MI: Prakken Publications, 1999.

Katz, Victor J. *A History of Mathematics: An Introduction*, 2nd edition. Reading, MA. Addison-Wesley Educational Publishers, 1998.

Klein, Aaron E. *Threads of Life: Genetics from Aristotle to DNA*. New York: Natural History Press, 1970.

Kline, Morris. *Mathematical Thought from Ancient to Modern Times*. New York: Oxford University Press, 1972.

Kowal, Charles T. *Asteroids: Their Nature and Utilization*, 2nd ed. New York: John Wiley & Sons, 1996.

Krebs, Robert E. *The History and Use of Our Earth's Elements: A Reference Guide*. Westport, CT: Greenwood Press, 1998.

———. *Scientific Laws, Principles and Theories: A Reference Guide*. Westport, CT: Greenwood Press, 2001.

Kutzbach, Gisela. *The Thermal Theory of Cyclones: A History of Meteorological Thought in the Nineteenth Century*. Boston, MA: American Meteorological Society, 1979.

Lapedes, Daniel N. (ed.). *McGraw-Hill Encyclopedia of the Geological Sciences*. New York: McGraw-Hill, 1978.

———. *McGraw-Hill Dictionary of Scientific and Technical Terms*, 2nd edition. New York: McGraw-Hill, 1978.

Laudan, Rachel. *From Mineralogy to Geology: Foundations of a Science 1650–1830*. Chicago: The University of Chicago Press, 1987.

Leake, Chauncey D. *An Historical Account of Pharmacology to the 20th Century*. Springfield, IL: Charles C Thomas Publishers, 1975.

Littmann, Mark. *Planets Beyond: Discovering the Outer Solar System*. New York: John Wiley & Sons, 1988.

Magner, Lois N. *A History of the Life Sciences*, 2nd ed. New York: Marcel Decker, 1994.

Margotta, Roberto. *The History of Medicine*. New York: Smithmark Publishers, 1996.

Mayr, Ernst. *The Growth of Biological Thought: Diversity, Evolution and Inheritance*. Cambridge, MA: Harvard University Press, 1982.

———. *One Long Argument: Charles Darwin and the Genesis of Modern Evolutionary Thought*. Cambridge, MA: Harvard University Press, 1991.

———. *This Is Biology: The Science of the Living World*. Cambridge, MA: Harvard University Press, 1997.

McGucken, William. *Nineteenth-Century Spectroscopy*. Baltimore, MD: Johns Hopkins Press, 1969.

Mendeleev, Dmitri. *Principles of Chemistry*. New York: P. F. Collier, 1901.

Meyer, Herbert, W. *A History of Electricity and Magnetism*. Cambridge, MA: MIT Press, 1971.

Moran, James. *Printing Presses: History and Development from the Fifteenth Century to Modern Times*. Berkeley: University of California Press, 1973.

Motz, Lloyd, and Jefferson Hane Weaver. *The Story of Mathematics*. New York. Plenum Press, 1993.

———. *The Story of Astronomy*. New York: Plenum Press, 1995.

Needham, Joseph. *A History of Embryology*. New York: Abelard-Shuman, 1959.

Nye, Mary Jo. *Before Big Science: The Pursuit of Modern Chemistry and Physics 1800–1940*. New York: Twayne Publishers, 1996.

Oldroyd, David R. *Thinking about the Earth: A History of Ideas in Geology*. Cambridge, MA: Harvard University Press, 1996.

———. *Sciences of the Earth: Studies in the History of Mineralogy and Geology*. Algershot, England: Ashgate Publishing, 1998.

Owen, Edgar Wesley. *Trek of the Oil Finders: A History of Exploration for Petroleum*. Tulsa, OK: The American Association of Petroleum Geologists, 1975.

Park, David. *The Fire within the Eye: A Historical Essay on the Nature and Meaning of Light*. Princeton: Princeton University Press, 1997.

Parkinson, Claire L. *Breakthroughs: A Chronology of Great Achievements in Science and Mathematics, 1200–1930*. Boston: G. K. Hall, 1985.

Parr, G., and O. H. Davie. *The Cathode Ray Tube and Its Applications*. New York: Reinhold Publishing, 1959.

Peebles, Curtis. *Asteroids: A History*. Washington, DC: Smithsonian Institution Press, 2000.

Pfeiffer, Carl J. *The Art and Practice of Western Medicine in the Early Nineteenth Century*. Jefferson, NC: McFarland, 1985.

Pinto-Correia, Clara. *The Ovary of Eve: Egg and Sperm and Preformation*. Chicago: The University of Chicago Press, 1997.

Purrington, Robert D. *Physics in the Nineteenth Century.* New Brunswick, NJ: Rutgers University Press, 1997.

Read, Oliver, and Walter L. Welch. *From Tin Foil to Stereo: Evolution of the Phonograph.* Indianapolis, IN: Howard W. Sams, 1976.

Rehrer, Darryl. The Typewriter. *Popular Mechanics* 173, no. 8 (1996): 56–59.

Robbins, Michael. *The Railway Age.* Manchester, England: Mandolin Press, 1998.

Robinson, Victor. *Victory over Pain: A History of Anes*thesia. New York: Henry Shuman, 1946.

Rocke, Alan J. *Chemical Atomism in the Nineteenth Century: From Dalton to Cannizzaro.* Columbus: Ohio State University Press, 1984.

———. Convention Versus Ontology in Nineteenth Century Organic Chemistry. In Traynham, James (ed.). *Essays on the History of Organic Chemistry.* Baton Rouge: Louisiana State University Press, 1987.

Ronan, Colin A. *Science: Its History and Development among the World's Cultures.* New York: Facts on File Publications, 1982.

Rudwick, Martin J. S. *The Meaning of Fossils: Episodes in the History of Paleontology.* Chicago: The University of Chicago Press, 1976.

Russell, C. A. *The History of Valency.* New York: Leicester University Press, 1971.

Russell, Peter J. *iGenetics.* San Francisco: Benjamin Cummings, 2002.

Serafini, Anthony. *The Epic History of Biology.* New York: Plenum Press, 1993.

Singer, Charles. *A Short History of Scientific Ideas to 1900.* Oxford, England: Oxford University Press, 1959.

———. *A History of Biology to about the Year 1900: A General Introduction to the Study of Living Things.* Ames: Iowa State University Press, 1989.

Spangenburg, Ray, and Diane K. Moser. *The History of Science in the Nineteenth Century.* New York: Facts on File, 1994.

Steele, Edward J., Robyn A. Lindley and Robert V. Blanden. *Lamarck's Signature: How Retrogenes Are Changing Darwin's Natural Selection Paradigm.* Reading, MA: Perseus Books, 1998.

Sterne, Harold, E. *A Catalogue of Nineteenth Century Printing Presses.* New Castle, DE: Oak Knoll Press, 2001.

Stringer, Christopher, and Clive Gamble. *In Search of the Neanderthals.* New York: Thames and Hudson, 1993.

Sturtevant, A. H. *A History of Genetics.* New York: Harper & Row Publishers, 1965.

Swetz, Frank J. *From Five Fingers to Infinity: A Journey through the History of Mathematics.* Chicago: Open Court Publishers, 1994.

Symons, Alan. *Nobel Laureates 1901–2000.* London, England: Polo Publishing, 2000.

Tattersall, Ian, and Jeffrey H. Schwartz. *Extinct Humans.* New York: Westview Press, 2000.

Taton, Rene (ed.). *History of Science: Science in the Nineteenth Century.* New York: Basic Books, 1965.

Thackray, Arnold. *John Dalton: Critical Assessments of His Life and Science.* Cambridge, England: Harvard University Press, 1972.

Travers, Bridget (ed.). *World of Scientific Discovery.* Detroit: Gale Research, 1994.

Trefil, James. Puzzling out Parallax. *Astronomy* 26, no. 9 (1998): 46–51.

Van Dulken, Stephen. *Inventing the 19th Century: 100 Inventions That Shaped the Victorian Age from Aspirin to the Zeppelin.* New York: New York University Press, 2001.

Varadarajan, V. S (ed.). *Algebra in Ancient and Modern Times.* Providence, RI. American Mathematical Society, 1998.

Walker, Peter, M. (ed.). *Chambers Dictionary of Science and Technology.* New York: Chambers Harrap Publishers, 1999.

Weeks, Mary E., and Henry M. Leicester. *Discovery of the Elements*, 7th edition. Easton, PA: Journal of Chemical Education, 1968.

Weissmann, Gerald. Aspirin. *Scientific American* 264(1): 84–90

Welch, Walter L., and Leah B. S. Burt. *From Tinfoil to Stereo: The Acoustic Years of the Recording Industry 1877–1929*. Gainesville: University Press of Florida, 1994.

Williams, James Thaxter. *The History of Weather*. Commack, NY: Nova Science Publishers, 1999.

Williams, L. Pearce. André-Marie Ampère. *Scientific American* 260(1): 90–97.

Wilson, Charles Morrow. *Trees & Test Tubes: The Story of Rubber*. New York: Henry Holt, 1943.

Windelspecht, Michael. *Groundbreaking Scientific Experiments, Discoveries and Inventions of the 17th Century*. Westport, CT: Greenwood Publishing, 2002.

Winston, Brian. *Media Technology and Society: A History: From the Telegraph to the Internet*. London, England: Routledge, 1998.

Woodward, Horace B. *History of Geology*. New York: Arno Press, 1978.

Young, David. *The Discovery of Evolution*. Cambridge, England: Cambridge University Press, 1992.

Internet Sources

Snider, Kellie Snisson. The History of Rubber, 2001. PageWise, Inc. (http://lala.essortment.com/historyofrubbe_rcml.htm)

The Nobel Prize Internet Archive. Ona Wu, 2001. (http://www.almaz.com/nobel/)

WebElements. Mark Winter, 2001. (http://www.webelements.com)

O'Conner, John J., and Edmund F. Robertson. 2002. The MacTutor History of Mathematics Archive (http://www-groups.dcs.st-and.ac.uk/~history/)

Encyclopedia.com, Tucows Inc., 2002. (www.encyclopedia.com)

Park, John L. 1996. History of the Cathode Ray Tube (http://dbhs.wvusd.k12.ca.us/AtomicStructure/Disc-of-Electron-History.html)

SUBJECT INDEX

The page numbers for subjects that are major entries in the text are shown in **boldface type**.

For publications, author names appear in parentheses.

NAME INDEX

ABOUT THE AUTHOR

MICHAEL WINDELSPECHT is Assistant Professor of Biology at Appalachian State University. He is the author *Groundbreaking Scientific Experiments, Inventions and Discoveries of the 17th Century* (Greenwood, 2002).